Social context and cognitive performance

Towards a social psychology of cognition

Jean-Marc Monteil and Pascal Huguet
Laboratoire de Psychologie Sociale
de la Cognition, associé au CNRS
UNIVERSITÉ BLAISE PASCAL, FRANCE

Routledge
Taylor & Francis Group

LONDON AND NEW YORK

ACKNOWLEDGEMENT

We would like to thank Caroline Barrett for the English translation of this book. We are especially thankful for her unfailing patience and support.

First published 1999 by Psychology Press

Published 2018 by Routledge
2 Park Square, Milton Park, Abingdon, Oxon OX14 4RN
52 Vanderbilt Avenue, New York, NY 10017

First issued in paperback 2018

Routledge is an imprint of the Taylor & Francis Group, an informa business

Copyright © 1999 Jean-Marc Monteil and Pascal Huguet

British Library Cataloguing in Publication Data

A catalogue record for this book is available from the British Library

Typeset by DP Photosetting, Aylesbury, Bucks.

ISSN 0892-7286
ISBN 13: 978-1-138-87374-2 (pbk)
ISBN 13: 978-0-86377-784-4 (hbk)

Contents

Preface

The social psychology of cognition investigates neither pure cognition nor social cognition. This ambitious book argues for the undeniable impact of the social context on individual thinking and cognitive performance. As such, it explains neither the isolated individual lost in thought, like Tolman's rat, nor the individual lost in images of other people. Not social in the outcome measures—though certainly social in their implications—this volume endorses social causes for cognition. Social contexts, people's personal history, comparison to others, anonymity or exposure, and the topic's status all alter supposedly stable abilities. Social context, or "social insertion", in Monteil and Huguet's terms, conditions cognitive performance, and with it, life chances.

With its focus on cognition in the context of social meaning, this book argues that people's autobiographies, expectations, and social experiences influence how they think, even under seemingly objective circumstances. As such, this work supports a movement to understand William James's prescription that "My thinking is first and last and always for my doing; and I can do only one thing at a time" (James, *Principles of psychology*, 1890). As has been argued elsewhere (Fiske, *Journal of Personality and Social Psychology*, 1992; *Annual Review of Psychology*, 1993), social psychologists worldwide have returned to the importance of "thinking for doing", as they record the impact of social goals on the direction of attention and therefore on social thoughts and, we see here, cognitive performance. Monteil and Huguet here propose that the consistency of one's current social context with one's previous experience can depress or facilitate school performance and that focused attention is the key.

The breadth of their theoretical argument, from cognitive, social, and educational psychology, promises to interest a range of readers. Their research programme will attract scientists seeking to understand cognitive abilities. Their findings will concern parents, educators, and citizens of every democratic society that believes in merit, assessed fairly and scientifically.

Susan T. Fiske
Distinguished University Professor
University of Massachusetts at Amherst, USA

European Monographs in Social Psychology
Sponsored by the European Association of Experimental Psychology

Series Editor:
Professor Rupert Brown, Institute of Applied and Social Psychology,
University of Kent, Canterbury, Kent CT2 7LZ

The aim of the series is to publish and promote the highest quality of
writing in European social psychology. The editor and the editorial board
encourage publications which approach social psychology from a wide
range of theoretical perspectives and whose content may be applied,
theoretical or empirical. The authors of books in this series should be
affiliated to institutions that are located in countries which would qualify
for membership of the Association. All books will be published in English,
and translations from other European languages are welcomed. Please
submit ideas and proposals for books in the series to Rupert Brown at the
above address.

Published

The Quantitative Analysis of Social Representations
Willem Doise, Alain Clemence, and Fabio Lorenzi-Cioldi

A Radical Dissonance Theory
Jean-Léon Beauvois and Robert-Vincent Joule

The Social Psychology of Collective Action
Caroline Kelly and Sara Breinlinger

The Social Psychology of Cognition
Jean-Marc Monteil and Pascal Huguet

Forthcoming Titles

Social Development and Adult Identity
John Bynner, Nick Emler, and David Romney

Attitudes of Mind: The Pragmatic Theory of Rational Cognition
Maria Lewicka and Karl Halvor Teigen

**Love, Money and Daily Routines: Joint Decision Making Under the
Common Roof**
Erich Kirchler

Stereotyping as Inductive Hypothesis Testing
Klaus Fiedler and Eva Walther

Introduction

Two basic principles usually inform scientific psychology's approach to the study of cognition. The first principle defines the existence of a subject endowed with universal cognitive properties and, simultaneously, of objects possessing intrinsic properties. The second principle asserts that individual cognitive productions and constructions are the result of the individual's application of his/her universal properties to the object's intrinsic characteristics. Hence, in the context of a psychology known as cognitive, the research procedure typically consists in setting up, either by simulation or experimentation, the conditions for observing the participant's implementation of the very properties used to process the object. In this context, the possible social dimension of the processing under observation is either totally neglected, or understood solely from the standpoint of the object's characteristics.

Based on nearly 20 years of research in the field of the social regulation of academic performances (Monteil, 1997), this book offers both theoretical and empirical arguments in favour of the inclusion of the social dimension of human beings at the very level of the processes in operation in their cognitive activities. Such a proposition does not lessen in any way the importance that should obviously be attached to the individual's cognitive properties as well as to the object's characteristics. However, it does imply a willingness to accept the fact that mental activities and their products may depend, at times to a large extent, on the link between the individual and the social context of cognition, which includes the object's characteristics.

Indeed, it appears that this connection is not the product of the situation *per se* but rather the fruit of a personal experience activated in memory following a relevant social integration. This fact alone justifies granting this experience an authentic psychological status as well as supporting the idea of the creation, under certain conditions, of "a cognitive context of the self" irreducible to the external context.

Despite a series of fairly homogeneous results, this book offers neither a thorough theory nor prescriptions for action. It does, however, point to research and intervention directions where the study of cognitive performances, and more largely the study of cognition, is likely to find in social psychology both theoretical and applied grounds to work on.

We will first devise a concept that aims at laying the foundations for a "social psychology of cognition". This approach suggests that the social dimension of human beings is no longer only a derived or reactive element of a separate human activity but represents a truly coexisting reality. For this reason, the status of the social in the modern approach to cognition will constitute the central question in Chapter 1.

Using the current literature on autobiographical memory and on social comparison, we will focus, in Chapter 2, on building upon the cognitive and social foundations previously laid down.

In the next three chapters (Chapters 3, 4 and 5), we will endeavour to construct an empirical basis for this same approach. Experimental results suggesting a close connection between events and academic contexts on the one hand, and cognitive activities and performances on the other hand, will be presented. By focusing attention on the involvement of the participants' past experiences in the determination of their current cognitive behaviours, these results will strengthen what constitutes most certainly the key hypothesis of this work, namely that *the social context only exists through the intervention of cognitive structures of contextualization, such as those related to the individuals' autobiographical memory*.

Fortified by these results and by their interpretation, we will question, in Chapters 6 and 7, the current systems of explanation concerning "social facilitation" and "social loafing" phenomena. Indeed, these phenomena, thoroughly studied and thus well documented in the international literature, are the expression of the influence of social situations (presence of others and collective work) on individual performances. Well-equipped with original experimental results on the matter, we will offer an understanding of the phenomena in question in the general context of the social regulation of cognitive performances and functionings.

Finally, enriched by the theoretical and experimental arguments presented all through this book, Chapter 8 will provide a few markers for action in the academic field. In this last chapter, we will focus our attention on the issues of classroom management and of its effects on academic performances.

More generally, the rhetorical organization of this work is by no means built on the argument of a research process revisited, and thus reconstructed. On the contrary, it is carefully worked out from the faithful, albeit summarized, expression of the chronological development of studies and experimental results, of their interpretation, of their theoretical effects and of their extension to a context which has become larger than the initial field.

You will thus put your footsteps in those of the researcher, you will be able to castigate his erring ways, disprove his easy certainties or even recast his interpretations. This manner of discussing research and its extensions according to temporal reality has no other motive than to share with you a cognition in the making.

CHAPTER ONE

Towards a social psychology of cognition

When scientific psychology approaches the study of cognition, either through the processes allowing its construction or through those involved in its mobilization, the prevailing theoretical paradigm is still that of cognitive psychology.

As in any scientific method, the aim is to interrelate theoretical models and empirical facts. This dialectic is expressed through the confrontation between two approaches. The first, which is essentially a modelling approach, shares with artificial intelligence an interest in simulating more or less complex behaviour. The second, which is an experimental approach, seeks to confirm its hypotheses through empiricism. No one will dispute the validity of these theoretical and methodological orientations. Both are essential and fruitful, and provide insight into human mental activity. Both state the problem of cognitive acquisition, construction, and production in the following terms: on the one hand, a subject with allegedly universal cognitive properties, and on the other hand, an object endowed with intrinsic properties. The individual is assumed to apply these universal properties to the intrinsic properties of the object. The research procedure consists in setting up, either by simulation or by experimentation, the necessary conditions for observing the participant's implementation of the properties used to process the object. The object may either be a person, a physical object or a problem to solve. In some cases, it is presented in various contexts in order to test for modifications in the way the processing is done by the participant. Can such a well-established conception undergo a significant change, or is it in a position to allow a parallel, perhaps even an alternative, conception?

The redefinition or recategorization of objects and issues previously considered well established often stimulates new research or revitalizes a field of study. Such transformations in the meaning of objects are likely to bring deep and lasting changes in every discipline. Indeed, they may either reduce or extend the field of study, highlight neglected objects, introduce new perspectives or modify prevailing theoretical conceptions. A process of this kind has been at the root of the most extensive modifications in many areas of social psychology in the course of the last fifty years.

HISTORICAL BACKGROUND

Such a renewal first took place at the end of the sixties and the beginning of the seventies, when Heider's (1958) primary work introduced the concept of attribution. Researchers then began to look at the way people in their daily lives assign causality to their own behaviour, to the behaviour of others, and to impersonal events occurring in their environment. Once stated, this issue generated a great deal of research.

With the emergence of the social cognition trend at the end of the seventies, the objects studied in social psychology underwent a new transformation. The quest then became the understanding of how people perceive the world and the social relationships assumed to be its foundations. With this trend (cf. e.g. Higgins & Bargh, 1987; Schneider, 1991; Sherman, Judd, & Park, 1989 for reviews), mainly derived from the theories and research paradigms of cognitive psychology, the question of attribution became just another issue in the field of social cognition. Supported by the information-processing metaphor, the popularity of this trend was to some extent a reflection of the growing influence of cognitive psychology. From then on social psychologists began to direct their research activity towards new objects: encoding, organization, storage, recall, or recognition of social information. These new directions raised issues that required the development of more vigorous and complex methodologies that could help provide insight into more subtle cognitive processes and mechanisms.

Social psychology—at least its experimental form—may be in fact considered cognitive since the forties. Asch's (1946) work on impression formation is still quoted. Despite this past, the social cognition movement in some sense reshaped the questions faced by social psychologists by drawing their attention to the work done in cognitive psychology. However, this increasing concern with intrapsychic processes became (and probably still is) the basis of some exceedingly vivid controversies about social cognition.

OF SOME CONTROVERSIES ABOUT
SOCIAL COGNITION

While some psychologists, such as Simon (1976), see no substantial difference between cognition and social cognition, others, like Zajonc (1989),
clearly differentiate between the two. The former consider every cognitive
phenomenon to be governed by the same general process. The latter think
that the characteristics of human beings, both as objects of knowledge and
as knowing individuals, make social cognition and perception qualitatively
distinct from the perception of inanimate objects.

The social cognition trend has been "criticized" in many other respects,
such as its immoderate use of information processing theory, its neglect of
affects, motivations, and other social "practices", its failure to consider the
context in which cognitive phenomena take place (social interactions,
group membership, norms and values, etc.). In short, the matter is still
hotly debated (Beauvois, Monteil, & Trognon, 1991; Leyens, Yzerbyt, &
Schadron, 1994), but also quite sterile.

We shall not get involved in a confrontation of arguments trying to
defend a conception that either integrates or differentiates cognition and
social cognition. Relying on the results of the experimental studies presented later in this book, we would merely like to put forward certain
reflections that will hopefully launch a debate in a less conventional conceptual area of social psychology.

Cognition and Social Cognition

In our opinion, the problem is not whether or not the structures and
contents of the knowledge involved in information processing are different
depending on the social or non-social nature of the information. Neither is
the question one of choosing between individual cognition and collective
cognition, i.e. between individual knowledge structure and contents, and
collective or shared knowledge structure and contents. On a functional
level, it is most important first to describe and understand how individuals
construct their cognitions through interaction with the environment, and
which mechanisms and skills they use to construct this cognitive world. At
this level, the distinction between cognition and social cognition is not very
meaningful. We must next describe and understand how the products of
this cognitive world are exchanged and shared, i.e. how they are
"socialized", and act in their turn upon the cognitive apparatus through a
kind of social constructivism. At this level, too, the preceding distinction is
somewhat meaningless. However, the social psychology of cognition can
find the bases for its existence in the conjunction of these two objectives.

In the academic context of cognitive psychology, it must be acknowledged that the social is, at best, considered to be a derived or reactive

element of human activities. We may notice that a large part of the cognitive social psychology inspired by Heider derives from a general paradigm of this type. The fact that the object (and therefore the information) is social does not separate cognitive and social psychology. It is simply conceded that the social dimension somewhat complicates the models likely to account for cognitive functioning.

Conversely, there also exists the socio-determinist approach, according to which cognitive competence and its development would only depend on influences from the environment and on the place held by the individual in social relationships. Both these positions, if taken as mutually exclusive, are obviously unsatisfactory for social psychologists. In our opinion, neither the intra-individual level nor the social positions and societal relationships level should be neglected. Indeed, both should be taken into account, which is precisely what social psychology is trying to articulate (Doise, 1982). But we will discuss this topic later. For now let us be satisfied with neurobiology's unexpected support of theses substantiating the importance of social experience in the individual's structuring (Monteil, 1993a).

THE SOCIAL DIMENSION OF THE "NEURAL INDIVIDUAL"

Amongst the most highly active and fruitful lines of research, those focusing on the construction of the brain are rapidly acquiring new knowledge. These studies inform us about the structure and functioning of an organ that, not so long ago, was thought to be an impenetrable black box. Moreover, some people consider that the rapid development of neurosciences, including the neurobiology of behaviour, marks a true intellectual revolution: the "neural revolution" (Changeux, 1983). As a body of research focusing on the structure, development, and behavioural effects of the central and peripheral nervous systems, the neurosciences have quite understandably fascinated psychologists. This fascination has sometimes led them to adopt the neurosciences' discursive reasoning, even though they often lacked the scientific expertise to do so. Consequently, rather than subjecting psychology to the neurosciences in order to give it a new legitimacy, our knowledge of human functioning would no doubt gain more if we retained only those contributions that would allow psychology to further its participation in the realm of the basic sciences, all the while maintaining its specificity. It seems to us that some of these contributions justify the usefulness of psychology, and more particularly of social psychology.

A Neural Postulate and Cognitive Approaches

No one today questions the fact that the development of the brain is the result of genetic programming. However, everyone also agrees that in the human species, more than in any other, the execution of that programme is quite flexible. This flexibility, called epigenesis, precludes discussion of a strict determinism: "Every human being has inscribed in the very structure of his brain through particular neural networks, the special affective, social and cultural history that is his" (Lecourt, 1989, p.142, our translation). Certain studies provide a remarkably striking illustration of this determining role of epigenesis. For instance, we may consider the studies on the Japanese brain that show hemispherical specialization for the use of two different writing systems. The alphabetical system, the Kana, relies on the left hemisphere, while the ideogrammatic system, the Kanji, relies on the right hemisphere. In human beings, this indeterminate part, linked to epigenesis, opens the possibility of acting upon the very programme that itself *is* transmitted. Indeed, the higher up the evolutionary ladder one moves, the more the epigenetic component gains importance in the construction of individuals. The more this epigenetic part is important, the closer the link between the structure of the nervous system and the individual's history: "Man's central nervous system forms a kind of engram of his personal history, and the human individual, unique and thus "unclonable", is the product of his social history ... This history is marked in the individual's physical structure, i.e. it is written in the cerebral matter itself, owing to the importance of epigenesis, which will stabilize one circuit or the other" (Prochiantz, 1989, p.18, our translation).

If, as suggested, the social dimension has an influence on the construction of neural networks to the extent of being undissociable from their development, strictly neuroscientific psychologists should find therein some matter for thought. Indeed, don't they postulate a one-to-one correspondence between each individual mental state (situated in time and space) and a specific neurophysiological state? Cognitive psychology, which to some extent still keeps its distance from neurosciences, views this identity relation as simply occasional, in which case we speak of functionalism. Such a conception allows cognitive psychology to reconcile two requirements: One is to be part of the natural sciences, by virtue of which it claims to share a materialistic or physicalistic ontology; the other is the necessity of maintaining a certain degree of explicative autonomy with regard to neuropsychology. Consequently, it is difficult to grasp the reasons why cognitive psychology would regard social dimensions as outside its realm, since their importance is underlined in the sciences that use them as a point of comparison, or even of reference. If physical matter bears the mark of the individual's social history, it becomes conceivable that a symbolic "engram" of the social dimension might exist in long-term memory and might play a part in the development and the cognitive functioning of the human being.

It thus seems difficult to exclude this dimension from the study of learning processes, particularly if one considers, as does cognitive psychology, that there exists a highly structured inner life and that explanations need to be sought on the basis of well-ordered sequences of inner states. Classical cognitivism indeed recognizes that propositional attitudes (i.e. mental attitudes related to propositional content, such as believing that, wishing that) are real states of organisms that determine behaviour. The importance of taking the subject's social history into account thus becomes apparent, since it is at the very root of the construction of propositional content. The research into cognitive explanations is also concerned with representations and mental states situated at infra-personal levels. Fodor (1983) has distinguished two cognitive systems, which are quite autonomous in certain respects. The first one, peripheral, is characterized by an isolated information processing system. The second one, central, allows for integrated inferences. The former refers to the infra-personal level of cognition, whereas the latter designates the personal level. In both cases the social cannot be dismissed. Indeed, in the peripheral system, the social appears as a component part of the stimulus—the percept—and in the central system it can participate in the development of the concept. The distinction between percept and concept must be understood here in terms of the meaning given by Woodfield (1986).

Another issue among others, although equally important, remains the existence or not of two types of thoughts: a first type of limited content, simply internal to the individual, and another type of wider content dependent upon the individual's environment, such as is exemplified by Putnam with his "la terre jumelle" thought experiment. This would entail that two individuals could have the same inner psychology even though their thoughts might have different referential contents. Going even further, we may also consider that every thought content may depend for its individuation on external factors, among them social factors. This externalistic view of cognition might question cognitive psychology in a strict sense, while maintaining the reality of mental attitudes. It is noteworthy, however, that Searle (1985) has disputed Putnam's "extensionalism".

Without going any further, and no matter the rivalry between the different approaches to cognition, it seems perfectly conceivable to view the human individual as a *socially inserted neurophysiological and pyschological system*. As such, the individual constructs mental representations, and processes and stores information to be later activated (either automatically or consciously), in order to act in the real world. Thus, if one truly wishes to understand human cognitive functioning, social insertions cannot be neglected. Because of the importance newly given to "social insertion", we will later on provide a precise definition of it.

SEVERAL SOCIAL PSYCHOLOGIES?

In 1989, the *European Journal of Social Psychology* (see vol. 19, no. 5) raised important epistemological questions, specifically in the context of the psychosocial study of human cognition. Indeed, these questions allow for a confrontation between various viewpoints, which in the end may lead, at least in our opinion, to a synthesis between the cognitive sciences and social psychology. The first viewpoint, defended by Zajonc (1989) and Nuttin (1989) and inspired by the natural sciences, sought explanations in terms of inner causal mechanisms in the individual. Conversely, one could also point to a social psychology rejecting any explanatory system based solely on inner mechanisms, offering instead a social constructionism approach (Gergen, 1989) wherein mental life takes its root essentially in the social. Close to this idea, Harré (1989) considers, moreover, that cognitive psychology, which studies mental life at the individual level, is nothing but an illustration of what he calls the "privatization of the social". Gergen and Harré's stance is seemingly close to Doise's (1989) and Moscovici's (1989) "psychological constructivism". However, according to Gergen and Harré, the fact of viewing the psychological as basically social leads them to reject the experimental methods championed by Zajonc and Nuttin, and to offer instead investigations using as a starting-point discourse analysis, for example. Doise and Moscovici's constructivism, on the other hand, relies on psycho-logical methodological approaches.

Without moving any further into this epistemological debate within social psychology (see notably Israël & Tajfel, 1972), a few basic problems concerning social psychologists' views on cognitive functioning may however be identified. Firstly, the status of the social within social psychology should be questioned, for it is a status that appears to be variously defined, to say the least. Secondly, one must note the lack of a thorough reflection on the status of the cognitive, especially since mental life and its products really *are* the centre of preoccupations in this area. In summarizing the various positions, two particular stances are capable of including all viewpoints and allowing us to enquire into the role of the social in the field of the psychology of cognition. In the first stance, the social is defined to a certain extent by the individual, and the psychological functioning finds its own material basis in the brain. The social is thus a derived and reactive product of an individual activity governed by physiological as well as neuro-physiological laws. Secondly, in Doise and Moscovici's stance, social interaction, communication, and social meta-systems, permeated by norms and values, influence cognitive functioning and participate in the construction of new forms of cognition. However, the mechanisms and processes through which the social informs and influences this functioning, thus participating in cognitive activity, still remain to be identified.

It is probably not enough to state, along with Vygotsky, that individuals internalize what they construct first in the social in order to be able to explain this internalization process. Likewise, one may conceive that human behavior can be directed by socially shared representations (Moscovici, 1961, 1984). But exactly how these representations are constructed and how they evolve in time and space remain relatively unclear (see Huguet & Latané, 1996; Huguet, Latané, & Bourgeois, 1998b). Transmitted and communicated by other individuals through everyday micro-contacts, such representations would appear to be changed into public representations by the person who communicates them, then changed again following a kind of epidemiological process (Sperber, 1985; 1996). Such an explanation, however, is not completely satisfying: We must still pursue our inquiry concerning the generation of the very first representation, thus bringing us back to the question of individual cognitive activity and of its possible social determinants.

These elements illustrate quite well a certain difficulty in truly grasping the social dimension following a common viewpoint and a univocal definition of its influence on cognition.

WHAT PLACE DOES THE SOCIAL DIMENSION HOLD IN THE STUDY OF HUMAN COGNITIVE FUNCTIONING?

If we agree that one of social psychology's specific aims is to understand the functioning of human beings as both participants in and partakers of the social world, we should easily subscribe to the following statement: Studying the individual from a psychosocial point of view entails first considering the individual as a socially inserted being, and then striving to construct explanatory systems that take social insertion into account.

Let us return for a moment to the neurosciences. Behavioural neurobiology teaches us that the social and affective history of individuals is engrammed in the cerebral matter itself. Yet this social history is nevertheless the result of a series of individual insertions that may vary in permanence and specificity, but that at any rate *do* exist. Consequently, it seems quite logical and acceptable to consider that if social and affective history finds a means of expression in the stabilization or in the form of particular neural networks, it also finds its symbolic expression in the human being's permanent memory. This leads us to formulate the hypothesis that, at the time of information encoding and storage, social and emotional experience is closely linked to the individual's corporal environment. There is good reason to believe that this hypothesis is by no means absurd. Referring to Tulving's work (1972; 1983), we know that what is stored corresponds not only to information *per se*, but also to the way in which that information was processed and placed in memory. Endowed with a representational content, social and emotional (probably interrelated) experience is to be located in the realm of mental representations. In agreement with Thomson and Janigian's suggestions (1988), human individuals cognitively represent their lives in terms of the events they experience, the goals they achieve, their failures, successes..., forming as it were a life schema that, like every schema, is assumed to act as a processing unit for incoming information. The notion of episodic memory is quite relevant here, for it reminds us that information is coded both temporally and in connection with the person actually doing the memory storing. It thus appears fair to think that one's personal history, which results from one's social history—because it depends on the corresponding social insertions—is strongly imprinted in memory. For this reason, this history is easily activated and implemented in the form of a system of responses found in the individual's behavioural repertoire. For this to happen, the individuals need only to find themselves in the presence of certain inputs or certain sociopsychological configurations acting as retrieval cues for knowledge related to previous social insertions. More of our attention should thus be directed towards contextual information of an episodic nature, especially in the contextualist trend of cognitive psychology. But, most probably for reasons of methodological convenience, contextual information of a semantic nature has been given greater importance. The difficulties encountered in taking this episodic dimension of human memory into account constitute a legitimate explanation of why cognitive psychology has left the social history of humans out of its field of research. In the study of the processes underlying the construction of knowledge, these difficulties have also led to viewing the social dimension merely as a source of external variation, at most liable to modulate behavioural responses, but never to generate them.

The use of the mnemonic system to understand and explain human behaviour has received substantial support, although it may still be viewed, along with the information processing linked to it, as socially inserted. This perspective inevitably leads us to take the individuals' social history into consideration in order to understand their cognitive behaviour and social conduct. Such a requirement has implications not only for the theoretical models used to support the social psychologist's scientific activity, but also for the experimental paradigms used with regard to the way in which variables are defined and manipulated.

OUTLINE OF A MODEL

Adopting the view that humans are socially inserted beings presupposes that all the consequences of that insertion should be considered in the study of cognitive activities. If we continue to agree with Tulving that what is stored in memory is not just information but also the way in which the information was processed and retained, we can reasonably assume that the subject encodes both the information and its context. In the concept of the inserted individual, the context of that information is mostly composed of the social and emotional conditions in which the subjects find themselves when faced with information to be processed, or with the necessity of forming an attitude, devising a strategy, or managing a mechanism to render coherent the environment. The idea that the context has an influence is not something new, if we consider the work initiated by Rosenzweig in 1933!

In short, in a learning situation, the subject is always faced with the reality of a social insertion that accompanies the information to be processed. The construction of a rule or of a piece of knowledge is thus specific to that insertion. What is imprinted in memory is a representation that associates the insertion (or some of its characteristics) with the rule or the knowledge (or some of their characteristics). We could consider the possibility of processing this representational content separately, but to do so we would have to acknowledge the total independence of the information to be processed from the whole of which it is part. This hypothesis presupposes that the individuals are capable of cutting themselves off from the environment, which is obviously difficult to carry out, on a neuro-physiological as well as on a psychological level. Does this mean that, for the individual, all information is always "ecologically" defined? Certainly, in our opinion, and for the very reason of the existence of an individual defined as socially inserted. The problem of knowing whether the information is processed first at the level of sensorial memory, that is to say, in the early stages, or whether it is indeed processed at the level of working memory does not make any difference to the basic fact that this information was encoded in a specific situation of insertion. One must therefore contemplate the fact that the knowledge, the rules acquired by an individual, and the processes that support this acquisition are constitutively dependent on the social conditions of their development.

And where does social psychology fit into all this, one might ask? Its existence is fully justified here, as it is a matter of giving a preponderant place to the social dimension of human beings in the construction and use of knowledge. When we say "construction" and "use" it should be understood that we are dealing with two distinct processes. The literature on knowledge in social psychology sometimes neglects the distinction between acquisition and retrieval of knowledge, particularly when speaking of a social psychology of cognitive constructions. It is indeed necessary to note that the processes of acquisition are in keeping with information to be encoded, knowledge to be constructed, and, therefore, that here the social dimension of which we speak is involved in and has something of the nature of this construction. The processes of retrieval and of mobilization reflect the attempts to establish a correspondence between current problem situations and formerly developed knowledge.

This distinction is important. It actually determines two research approaches, complementary certainly, but, nevertheless, heuristically separate. One aims to understand how, and by what mechanism(s), the social dimension constitutively affects cognitive constructions; the other focuses on two things: (1) to understand the activity of cognitive production by studying the modalities by which correspondences are made between current situations in which the individual is inserted and those that were constructed and stored at a certain time in the history of this same individual; and (2) to identify and understand the activities brought into play when this correspondence is not established.

These two approaches have, of course, methodological implications that bring about strong constraints, notably at the level of experimental work. For instance, experimentally proving the influence of social insertion on the development of rules or of knowledge presupposes first manipulating and creating the social conditions of acquisition of these rules and knowledge and then verifying that the use of the latter for production purposes finds expression, all things being equal, in differences of attitude, strategy, or performance, according to the manipulations carried out. In so far as empirical elements would appear to indicate the presence of social conditions in the experimentally controlled development of this cognitive processing/"schema", one would dispose of important information for undertaking research focusing strictly on the mechanisms by which this exertion of control—in the cognitive sense of the term—of social conditions takes place.

The notion, defended here, according to which all processing of information would take place under social insertion, raises, at the same time, the question of the link between "contextual" information of an episodic nature and "contextual" information of a semantic nature. Can one actually consider these two "contextual" structures, episodic and semantic, as dissociated or dissociable? The question remains open.

Whatever the answer to that question may be, the general perspective adopted here implies also contemplating experimental paradigms that, in the first place, allow one to control the social conditions of the development of knowledge and then expose the subjects to scenes capable or incapable of entering into correspondence with what was previously constructed. It is easy to see where the difficulty lies in such a programme. It necessitates the processing by the individual of information in an experimental situation such that its effects at the level of cognitive activity, realized in situations that are either incongruent or congruent with the manipulated insertions, can be measured.

To help understand more precisely human cognitive activities, it is not enough for social psychology to study the behaviour of individuals and groups in social contexts. Indeed, a problem at once arises. Must one tackle the study in itself of the social context of cognition or rather approach it through the possible social determinants of its development, of its mechanisms, or of its forms of expression? The two things are indeed quite different. In the first case, the problem raised would rather suggest a theoretical conception of an ecological type (Gibson, 1979) at the heart of which the behaviourists' stimuli would have gained the status of events. These events, even if they are considered as contexts for understanding and action (Schanck & Abelson, 1977), or as settings making the recall easier (Neisser, 1986; Nelson, 1986), are nevertheless difficult to delimit (Avrahami & Kareev, 1994). At times defined by the objects on which they are based (Quine, 1985), by the causes that generate them and their effects, or even by the goals guiding the action, their limits remain hard to grasp. Nevertheless, for this type of approach, the information is not in the mind but in the world, and presents itself directly to the perceptual system through the ecological coherence of events. This is obviously a conception critical of the dominant cognitivism, and close to a contextualist behaviourism (Hoffman & Nead, 1983).

In the second case, as implicit as it may be, the criticism of cognitive psychology is not absent but rests instead on other bases. By explicitly assigning a role to the social context in the approach to cognition, this conception in a way conflicts with Minsky's (1985) famous "society of mind". The representation of an individual cannot by any means be reduced to that of a solitary information-processing machine. The fact that human cognition takes place in a cultural and social universe (Bruner, 1991) and that it expresses the result of both the processing and the integration into an environment presenting properties that the individuals place in memory along with the meanings that ensue from the nature of their social insertions (Monteil, 1993a) should thus not be neglected. To take into account and pay attention to the individual's possible motivations, as well as to his/her interactions, real or symbolic, with others, or to consider his/her past experiences, thus becomes an epistemological requirement. In such a conception, it is neither the event's ecological coherence nor the transaction between its cognitive properties and the objective properties of such and such a task alone that guides and determines the individual's performances and actions; rather the guiding principle may be found in the social link concerning this same task or this same object in which the individuals find themselves involved.

It is thus indeed from the vantage point of a social psychology of cognition (Monteil, 1993b), not from the angle of a psycho-ecology, of a neo-behaviourist contextualism, or even of a microsociology à la Goffman (1959, 1967), that one should approach the issue of the social context and deal with its possible influence on individual cognitive constructions and performances.

The Notion of Context

Science, having for a long time neglected or even rejected problems related to context, has finally acknowledged that a certain number of effects can indeed be attributed to it. Physics, particularly mechanics, was intent on eliminating any outside influence on phenomena examined, be it the world's influence or the observer's. The other sciences yielded to this "rule" separating the essential from the unessential. Human sciences, at first taxonomist, eliminated the issues raised by the interrelatedness of human conducts linked to the contexts of their inscription. Experimental psychology, including social psychology, did not elude this principle, dogmatic behaviourism being its driving force. Indeed, for behaviourists, any response of the organism is determined and controlled by a specific stimulus. Consequently, the stimulus is an efficient part of the stimulations stemming from the environment.

Physics had disposed of the questions raised by the context. By a seeming paradox, it was also from physics that a renewed interest in taking them into consideration was to originate. At the beginning of the century, the physicist Mach's criticism of Newtonian mechanics and Einstein and his theory of relativity were to promote the idea that there is no such thing as an isolated event. At that time too, the Gestalt theory modified scientific psychology, with, for instance, Wertheimer and the perception of movement and Köhler and the intelligence of apes.

Up until this point, however, the contexts considered are essentially those defined by the outside environment. The influence of the history of the process on itself is still neglected. Wiener's (1948) cybernetic conception re-established the idea according to which the history and the effects of a process may have an influence on its evolution and on its properties, the retroaction principle associating a dynamic system to its previous states, and thus to its own context, being at play here.

In the field of psychology, Tulving was to play a determining role with his conception of an interaction between memory encoding and retrieval, and his formulation of the specific encoding principle. Indeed, to formulate the idea that the probability of retrieval of a piece of information depends on the compatibility between the contexts in which the individual encodes and retrieves this information is to place the context at the very heart of the cognitive processes. In the same way, to distinguish between, on one hand, episodic memory attached to personally experienced, localized, and dated events, and on the other hand, semantic memory related to facts and concepts, is to give personal experience a fundamental psychological status and, in a certain way, to mark the existence of a personal, inner context (the cognitive context of the self) against an "external" context (see also Chapter 8).

However, one of the main issues raised by the context remains that of its definition (Tiberghien, 1986). Indeed, how can one identify the context, in so far as any event can be taken as a part of it? It thus seems that we are not actually faced with an object, of whatever nature, but indeed a process, a *contextualization process* as it were, which may manifest itself in various situations and objects. In a Tulvingian conception of memory, the codings experienced by the cognitive system are thought to be stored in memory, and the process allowing access to them is considered to result from the interaction between a retrieval context and a state of the system in memory recapitulating its history. In other words, meaning would emerge from memory's episodic functioning rules (Hintzman, 1986). Following this theoretical perspective, if we care to contemplate for a moment the view that the human individual's characteristic is to be socially inserted, some of the cognitive activities may then depend on the interaction between the cognitive system's history, its contents, and the spontaneous (or even current) conditions of their retrieval.

At this stage of our reasoning, we should point out the basically psychological definition that we have given of social insertion in the past (Monteil, 1993a). It means that a social position or condition has become dynamic by means of a modification in the place or situation of the individual. This now dynamic position or social state (i.e. a position or state that has now been changed into a social insertion), becomes a psychological event or episode. As such, it is encoded and stored in memory to be later mobilized, should the conditions for its activation arise. The salience of a social insertion varies, depending not only on the structure and features of the situation in which it takes place, but also on the events or sequence of events it may activate in long-term memory.

In short, in this perspective, conceiving humans as socially inserted beings means understanding them as historical individuals, i.e. as people endowed with memory. But the content of human memory does not result only from the processing of the intrinsic properties of objects, whether those objects are social or not. It is first and foremost the product of the relationships individuals have with objects, which in turn depend on their social insertions.

Individuals, indeed, are both actees and actors. The actions, options, strategies in which they are involved do not include in themselves their reasons and explanations. In other words, no one makes choices freely and knowingly. However, neither are these choices directly imposed by the outside world through the positions that the individuals occupy in the different social fields. Rather, they are directed from the inside by their socially constructed and deeply rooted "experiences" of cognitive capacities, of world vision and division, of hierarchy, of preferences. The links between the objective structures (those of the social contexts) and the incorporated structures (those of the individual's memory system) constitute in a way the basis of the individual's relationship to the object of knowledge—the latter, dependent on the former, being at the same time able to participate in their implementation as well as in their reproduction.

If social psychology is to participate in the study of the human cognitive system at all, then it must deal with a socially inserted system. The study of cognition cannot be envisaged without taking stock of the contents and the places upon which cognitive processes are based and in which they occur. These places are not to be understood as physical locations, but as psychological spaces through which individuals experience their social insertions in the human community. Accordingly, social comparisons, sociocategorical membership, interpersonal or intergroup situations are all melting-pots of these experiences.

The variety of these contexts and the difficulty of analysing them thoroughly must not be an excuse for neglecting them altogether. If social psychology should obviously have something to say about human behaviour that goes beyond the features of the contexts, it cannot for all that pretend to ignore them. To consider the preconceptions related to a single culture as universal is a pitfall that must be closely assessed; all the more so in that the results of experimental research do not come from *one* culture, but from a subcategory of that culture.

Aware of these limits, how must one understand and define social context in the field of scientific research in psychology? How can one discern social psychology's place in the current cognitive symphony? Which particular problems in social psychology would appear to be good examples of general interest issues for psychology, while still leaving social psychology its specificity as scientific psychology? All of these questions in social psychology underlie our efforts to define a psychosocial perspective in the study of cognition.

Cognition does indeed cover a large domain of studies and objects of study: judgements, inferences, attributions, perceptions, attitudes, ideologies, representations, goals, intentions, values, etc. It may also mean a reality shared by given communities, for instance the scientific community. In that way, it is also interpersonal and social, while subsuming intrapersonal cognitions. Moreover, intrapersonal cognitions are, in a way, social too. Indeed, many individual cognitions are acquired through our relationships to others by confronting our beliefs and views of the world with those of others. The facts of our experience are also social; therefore they depend, for their meaning and their organization, on concepts and categories previously constructed.

The social psychology of cognition should thus endeavour to study contents and structures of knowledge as well as the social and cognitive processes through which this very knowledge is formed and transformed.

Allowing for a wider scope than the social cognition trend, it would thus concern itself also with the processes supporting mental activity as well as with the contents and situations in which these processes and contents are at work.

Work on the processes is obviously fundamental, because it allows us to uncover trans-historical and trans-cultural regularities, and thus to strive towards the setting-up of the laws specific to the human species. Work on the contents is no less important, as it may allow us to construct the cognitive maps of individuals and groups, helping us better to understand reciprocal influences.

Work on social situations enables us to define and to grasp the links between the individuals and their objects of knowledge, whatever those objects may be. Because the social psychology of cognition can deal with these three levels of study, underscoring their otherwise unnoticed connections, it is bound to offer useful contributions to scientific psychology. Thus, even if it adopts the prevailing information-processing theory, it may put forward an individual vested with his/her social insertions who, like the Heiderian perceiver, gazes at the universe, but also acts on the world as much as he is acted upon by it.

The classical trend in social cognition may lead one to raise the question: What might be the "social" in social cognition? We may wonder if memorizing the characteristics of other individuals or judging a person or a group is very different from memorizing the characteristics of Kandinsky's or Klee's paintings or judging the similarities and resemblances between a set of *objets d'art*.

These questions have led Tajfel (1981) to call for a more "social" social cognition. The perspective drawn here does not raise the question of figuring out whether such and such a piece of information is more or less social and, thus, whether we are dealing with cognition or social cognition. The conception of a socially inserted person involves considering the relationships individuals have with the objects around them. These relationships are mainly determined by each individual's current social insertion and by the representations of prior activated or to-be-activated insertions. In other words, the human cognitive universe is the outcome of the processing and integration of an environment whose properties individuals store in memory along with the meanings ensuing from the nature of their social insertions.

Thus may we join together two different conceptions of the place of social context in psychology. The first conception adopts the point of view of the individual as "perceiver" or "information processor", interpreting the information provided by the social context, considered as a configuration of stimuli judged, interpreted, and memorized in the same way as any other configuration. The other conception views the human being as a participant in the social context and influencing it, either as an individual or as a member of a group. The latter conception leads, on the one hand, to viewing individual experience as a social product (we feel and we think in such a way because we are social beings) and, on the other hand, to viewing the social context as a product of action and of human thought (the world in which we live is in part the product of the way we live).

It is obvious that the first conception naturally provides a link with cognitive psychology, whereas the second more readily points to a coming together with social sciences such as sociology and economics, for example. It is not so much a question, however, of a social psychology of cognition's testing itself with regard to such and such scientific partners, as of drawing from them information likely to nourish its own approach. Neither a median discipline between cognitive psychology and the sociology of cognition, nor a social appendix of psychology, nor even a psychological outgrowth of sociology, it is beyond doubt that social psychology can keep and even strengthen its scientific identity through the links between the cognitive and the social.

As a matter of fact, certain studies clearly indicate the interrelatedness of these two fields. Voss, Greene, Post, and Penner (1983) have shown how participants from different professional backgrounds process the same problem-situation. They have noticed that the cognitive strategies used express, beyond any expertise, the very structure of the professional environment. Such studies have led Simon to write: "...since the performance depends heavily on socially structured and socially acquired knowledge, it must pay constant attention to the social environment of cognition" (Simon, 1990, p.16).

Other types of research are also capable of supporting an argument in the same direction, for example Doise and Mugny's work on social marking (1981): "Social marking exists when there is an homology between, on the one hand, the social relations defining the actors in a specific interaction and, on the other hand, the cognitive relations concerning certain properties of the objects that mediate these social relations" (p. 42, our translation).

CONCLUSION

The social psychology of cognition is neither required to merge with cognitive psychology, nor must it become a micro-sociology. Its area of study is the socially inserted individual, who maintains social contacts, participates in interactions, belongs to social groups, and is the object of categorizations. In this perspective, it is difficult to imagine that this set of characteristics can be taken to be negligible in the study of his/her cognitive functioning. It is therefore at a dual level of investigation of human functioning that social psychology must assume its scientific responsibility.

It must continue to devote itself to studying the effects on human behaviour of fundamental social situations, such as the presence of others (see Chapters 6 and 7), situations of social comparison, inter-group relationships, etc., so that one can better identify the most basic and most cognitively salient social insertions for all human individuals.

Fortified by this knowledge of the relationships between the individual and the collective, social psychology must then also promote and conduct a procedure aimed at grasping the study of human cognition by taking into account the autobiographical dimension of individuals, i.e. their social and affective history, which is a product of the insertions offered or imposed by everyday life.

The aim of all this is to determine to what point human cognitive operations can be socially regulated. This would show how individuation is also the expression and the result of the processing and the integration of the social environment stored in memory by means, particularly, of the psychological and neurophysiological competences specific to the human species.

It will be acknowledged that, since these concerns are not necessarily of interest to other fields of scientific psychology, the social psychology of cognition can therefore draw from them reasons for asserting its own unique contribution.

CHAPTER TWO

The cognitive and social bases of "insertion"

COGNITIVE BASES

To view individuals as socially inserted beings leads on to defining them as historical individuals; this is by no means an innovative idea. Indeed, temporal evolution inevitably inscribes us in a social history that also includes our own history. As a product of our successive experiences, this personal history probably forms an important part of our memory's heritage, thus representing knowledge that may be activated whenever the proper conditions for retrieval are met. As we have already mentioned, the memory of these personal experiences refers quite precisely to what Tulving (1972) calls episodic memory: "a system that receives and stores information about temporally dated episodes or events, and temporal-spatial relations among them" (p.386). However, in line with Larsen's definition (1985), we would like to draw a distinction neglected by Tulving in his own differentiation between semantic and episodic memory and their related knowledge, knowledge that can be described as con-textualized (episodic knowledge) and as decontextualized (semantic knowledge).

Since we intend to outline a social psychology of cognition, it seems useful here to distinguish between two dimensions of knowledge: a personal dimension and a situational dimension. The former includes a reference to the individual possessing the knowledge and the latter, a reference to a specific situation or to an episode; their junction thus providing autobiographical knowledge as understood by Tulving. However, by merging the personal and the situational dimensions, Tulving has neglected other categories of knowledge, which he considered to belong to the semantic memory, under the label of general knowledge. Such is the case for historical and factual knowledge and for self-referencing knowledge.

If one truly wants to consider facts, one will notice that they are always contingent upon a particular historical situation, whether the individual is aware of it or not. If this contingency did not exist, the factual knowledge would then be thought of as a general law and thus become a conceptual, entirely semantic, knowledge. In the same way, little attention has been given to knowledge connected with general impressions (non-episodic) and to beliefs concerning the self (our attitudes and preferences, our abilities, our shortcomings, our personal identity as a whole). To a large extent, this knowledge, generalized from episodic experiences, has nonetheless preserved the reference to oneself thought of as an object of knowledge. To put it plainly, alongside strictly episodic and autobiographical knowledge, these two last categories constitute knowledge that should indeed be related to the self.

With the reference to time or to the position of the event, we are clearly dealing with historical or factual knowledge. Indeed, chronological time is a prerequisite of all events, no matter whether they are found to be in a causal, an intentional, or an accidental relationship. Time is fundamental to the organization as well as to the coherence of the event (see the studies on retrograde amnesia, for example Baddeley, 1992, for a review). It is usually not cognitively represented as a calendar time, but rather in a chronological order of experiences defining the individuals' knowledge about their own history and about the world around them. Thus, this knowledge truly *is* contextualized.

The general self-knowledge (our attitudes, our shortcomings, our behaviour), even though not situated in time and space and not strictly autobiographical, is nevertheless knowledge of the self and about the self that can be activated (and probably is), and is perhaps even applicable and applied in some situations. Obviously, this knowledge, neither situated nor specific, nevertheless defines a general internal context about oneself.

These few elements of clarification allow us to put forward the idea that the carrying out of a given task, according to its conditions and, in particular, its social conditions, may bring about the involvement, either by explicit or implicit retrieval, of various categories of knowledge directly or indirectly related to the self (see Chapters 3–5 for empirical illustrations). Therefore, social insertion's main cognitive bases are likely to be found in the contents of memory directly or indirectly related to the self through the individual's experiences.

This proposition leads us to question the type of memory most closely linked to contents referring to the self, that is to say autobiographical memory. Indeed, as is emphasized by Rubin (1982), autobiographical memory is currently one of the themes of research authorizing and even requiring the most important articulations between various fields of study such as cognitive psychology, social psychology, developmental psychology and clinical psychology. Autobiographical memory calls for an integration of knowledge concerning the basic processes of human memory and the influence of society on individuals, as well as of knowledge about cognition and affect (Rubin, 1986). Without opening here a historical chapter, we may however note that Freud (1959) had already pointed to the importance of emotions (which we will discuss later) and of memories for subsequent behaviour.

Autobiographical Memory

In a very general sense, the term "autobiographical memory" encompasses a person's capacity to recollect his or her own past experiences (Baddeley, 1992; Robinson, 1986), or to record personal events (Neisser, 1986). This does not mean, however, that all memories must be immediately linked to the autobiographical dimension. In general, the self-referencing aspect of memories will determine their autobiographical aspect (Baddeley, 1992; Brewer, 1986; Neisser, 1986; Robinson & Swanson, 1990). Thus, we may assume that autobiographical knowledge implies a more important reference to the self than do other types of knowledge. Remembering a personal event in a particular spatio-temporal context would be an indication of the memory's autobiographical aspect. Because they refer to events with personal implications, memories and autobiographical facts are likely to be retrieved many years after the event's actual encoding. They also refer to complex representations, including sensorial information, as well as information pertaining to the individual's personal interpretation (Conway, 1990a). This last aspect renders the autobiographical memory closer to the beliefs and understanding of the rememberer than to the truth of the event. Thus, according to us, memory is truly the product of a social insertion. Following this line of reasoning, we may concur with Baddeley (1992), who differentiates autobiographical knowledge from other types of knowledge by distinguishing between the self as experiencer and the self as the object of an experience. This last characteristic is the only one likely to keep us from systematically considering each and every lived experience as autobiographical, and allows us to restrict autobiographical knowledge to that which, at one time or another, has engaged the self in the very content of the memory. Tulving (1983) underlines the fact that the knowledge stored in the episodic memory concerns events stemming from the individuals' personal past and, consequently, provides a basis for the definition of their personal identity. Conversely, the knowledge stored in the semantic memory is not necessarily linked to the individual's personal identity. In fact, Tulving views the terms "episodic memory" and "auto-biographical memory" as equivalent, which has even led some authors to substitute the second term for the first (Jacoby & Dallas, 1981). However, for various reasons, the supposed equivalence between autobiographical memory and episodic memory is far from being unanimously accepted (see Baddeley, 1992; Brewer, 1986; Conway, 1990a, 1991). Firstly, the method for approaching episodic memory, mostly centred on verbal acquisition tasks, discards the individual's personal involvement and thus his/her strictly autobiographical dimension. Secondly, autobiographical memory comprises not only elements of an episodic type but also semantic elements, as well as elements on the borderline between episodic and semantic, thus referring back to contents extending far beyond those retained by Tulving in his work on episodic memory. Thus, if episodic memory may indeed be considered as a system functionally distinct from other types of memory, and most particularly from the semantic memory system, this does not hold in the case of autobiographical memory, which contains at the same time episodic and semantic elements (Conway, 1990a).

Whether or not one wishes to make a clear distinction between auto-biographical memory and episodic memory (Brewer, 1986), whether one considers episodic memory as a subdivision of autobiographical memory (Conway, 1990a, 1991), or conversely autobiographical memory as a sub-category of episodic memory (Larsen, 1992), it still remains the case that autobiographical memory is not the equivalent of episodic memory. Rather, autobiographical memory is a memory process that stores and retrieves knowledge about the self (Baddeley, 1992).

Social insertion's cognitive basis would thus be a set of self-referencing knowledge constituting various areas all included in a system of episodic and semantic memory. Such an idea suggests that autobiographical memory should obey the same rules and regulations as the rest of the memory (Holland, 1992).

The original feature of the studies on autobiographical memory is to take into account the personal meaning of events, their emotive power (see Conway, 1991). Beyond its involvement in cognitive activities with a high level of integration (judgement, perception of others, problem-solving, etc.) and through its participation in the establishment and preservation of the links between individual histories, autobiographical memory should no longer be considered on a simple archival basis. As is underlined by Robinson and Swanson (1990): "Autobiographical memory is not just an archive: it is a resource for living" (p. 332; see also Barclay & Hodges, 1990).

To remember past experiences is a distinctive form of consciousness, suggested Tulving (1983). The personal quality of memories is one of the basic phenomenological traits of autobiographical recall. This quality reflects the extent of the link between recalled memories and a current situation organizing a personal experience.

But in which contexts and environments does this autobiographical recollection appear? One of the problems commonly facing individuals is that of predicting and interpreting the behaviour of others and adjusting their conduct according to the results of their analysis. Our ability to symbolize experience, to recall and use it for this type of activity, is increased by our very awareness of it (Humphrey, 1986). Therefore, this "self-consciousness" may support a subsequent process by providing everyone with a parallel model of everyone else's inner life. In other words, we use our own experience to generate hypotheses about the reasons liable to explain the actions of others, or perhaps even our own actions. For this reason, we may be able to construct the elements of an internal model of the functioning of others based on the autobiographical memory's functions. Moreover, the studies supporting a functional per-spective on autobiographical memory (Neisser, 1988), and according to which social development, development of the memory and inferential thought about others are interrelated, allow to link our own awareness of our subjective states of mind with our insight into others' feelings.

We may no doubt find here the elements for a renewed problematics concerning the perception and evaluation of others. Our autobiographical memories are probably not extraneous to the apparently biased form of our attributions and other daily explanations. Indeed, in order to analyse the situations in which we find ourselves, to make decisions, or to under-stand, evaluate and predict the behaviour of others, everyday life often leads us to refer to these memories.

Because, according to Reiser, Black, and Abelson (1985), cognitions are not only based on abstract generalizations about personal experiences, but also on multiple individual events encoded in memory, thinking of a given episode is also an essential component in the bringing into play of several cognitive processes such as those involved in learning and in planning, as well as in problem-solving (Reiser, Black & Kalamarides, 1986).

These various perspectives generate many questions: Which compe-tences are required to use personal experience best in the achievement of such goals? How are these competences acquired? So many questions lead today to just as many fields of exploration.

The explicit retrieval of personal memories obviously confirms the active participation of autobiographical memory in the processing of the miscellaneous situations encountered by an individual. However, this participation is probably also brought about outside the individual's control. Implicit retrieval should thus play a notable role in his or her activities.

The Implicit Involvement of Autobiographical Memory

Attention and Autobiographical Memory. According to Nielsen and Sarason (1981), the individuals' past experiences and the situation in which they find themselves may provide the contents as well as the organization of long-term memory, and would thus be able to determine in which way attention is allotted to different types of information. A schematic organization could regulate this attentional phenomenon. As is suggested in the literature (see Barclay and Subramanian, 1987; Robinson and Swanson, 1989), autobiographical memory would thus take the form of self-schemas (Markus, 1977), that is, cognitive structures recording generic knowledge about the self (stemming from past experiences) that organize and direct the processing of the information relating to the self. This information would encompass cognitive representations derived from specific events involving the person, as well as more general representations stemming from repeated categorizations and from either the evaluations of others about the individual's behaviour or the individual's own self-evaluation. Once created, a self-schema may play a mediating role in perception, memory, and action (see Brewer, 1986; Markus, 1977; Robinson & Swanson, 1990). In this hypothetical framework, Markus (1977) has suggested that, once established, self-schemas act like selective mechanisms determining the amount of attention that should be granted to the information. The relevance of this hypothesis was already present in Moray's (1962) famous experiment showing that the individual's own name broke through the barrier of a dichotic listening test. Bargh (1982) later contributed a more direct experimental proof of Markus' hypothesis. With the help of a dichotic listening test and with the double-task paradigm, he studied the effect of a self-schema on selective attention and showed that having an independence schema reduced the attentional effort in the processing of adjectives related to independence presented in the target listening channel. Conversely, if the adjectives connoting independence were presented in the listening channel to be masked, this same schema increased the processing's attentional effort. Bargh explains these results by suggesting that a self-schema supports a variety of long-term expectations constituting so many relatively permanent characteristics of the perceptual system. These characteristics are liable to arouse automatic attentional responses (non-intentional) to certain stimuli. Geller and Shaver (1976) have apparently confirmed the irrepressible nature of the processing of self-related stimuli. Using the well-known Stroop test (1935), they obtained, in conditions favouring attentional self-awareness, a more important interference effect with words referring to a self-schema than with more neutral words. By analogy with the studies on impression formation, where individuals seem to direct their attention according to the correspondence between their expectations and the target person's behaviour (see Hilton, Klein & Von Hippel, 1991; White & Carlston, 1983), we might contemplate the possibility that events inconsistent with a generic personal memory should call for a more important allocation of general resources than would events consistent with the memory.

In short, if the self is comparable to an autobiographical structure, then certain phenomena of the allocation and use of attentional resources are liable to be related to it. Thus, as is suggested by the empirical illustrations in Chapters 3 and 4, we must consider the influence of autobiographical contents on cognitive task performances.

Emotional States and Autobiographical Memory.   An account here of a few studies (Higgins, 1987, 1989; Strauman, 1990, 1992; Strauman & Higgins, 1987) will suffice to illustrate the links between self-knowledge and emotional states. According to the self-discrepancy theory (Higgins, 1987), the contradictions, or discrepancies, between what an individual would like to be or thinks one ought to be and his or her own self-concept represent negative psychological situations related to certain types of emotional discomfort. Strauman and Higgins (1987) have shown that, during a task unrelated to the self and requiring that the participant think in another person's terms, the discrepancies between an individual's standards and his or her self-concept were implicitly activated by merely presenting contextual indices such as words. Moreover, their automatic activation would bring about a negative emotional state. Still according to the self-discrepancy theory, self-standards appear to have been internalized during childhood following the acquisition of knowledge about the self and about others. Won over by this developmental perspective, Strauman (1990) forms the hypothesis that there exists a close link between the individual's standards and emotionally connoted childhood memories. He succeeds in showing that indices relative to self-standards (and particularly indices differing with the self-concept) make the retrieval of negatively connoted autobiographical memories easier, even in the case where these indices are allocated a positive valence.

Even if the contextual indices used here are of a semantic type, episodic contextual indices are no doubt able to instigate similar retrievals. Activated by such-and-such a situation, the knowledge stored in the autobiographical memory could thus induce in the individual the very emotional state to which it is associated. In fact, many studies put forward an important link between emotion and attention.

Emotional States and Attention.   The selective processing of a given piece of information and the accomplishment of a mental effort, both activities generally associated in everyday language with attention, are at the root of two fields of research (Johnston & Dark, 1986). The first one assigns a function of selection to attention, and applies itself to the dif-ferential processing of simultaneous sources of information, either internal (memories, knowledge) or external (objects, events). The second one, born in the seventies from the transition from a conception of the mind defined in structural terms to a more flexible functional notion (Kahneman, 1973; Posner & Snyder, 1975), no longer allots a function of selection to atten-tion, but rather a function of allocation of the available working ability. Also referred to in terms of "effort" and "attentional resources", this ability corresponds to a mental energy thought to be limited to the level of a central mechanism. At first considered as unique and non-specific (Kahneman, 1973), the storing of these attentional resources was after-wards thought of as plural (Navon & Gopher, 1979; Norman & Bobrow, 1975), mostly because different interferences were observed according to the tasks processed by the participants.

The limited-capacity models conceive attention as controlling, on the one hand, the amount of energy assigned to performing a given task and, on the other hand, the distribution of this energy among the various activities with regard to their characteristics and to the order of priority granted to them by the individual (Richard, 1980). Thus Kahneman (1973) suggested a variable allocation of capacity model. This type of model is based on an empirical proposition according to which performances on two tasks simultaneously carried out may be manipulated in such a way that, with regard to their relative difficulty, one may be better done than the other. Therefore, this attentional capacity would thus be made subject to a division but also to an allocation assigning, according to Kantowitz (1985), the application of capacity to a particular task and the attribution of resources to a specific processing of the task. However, since, according to Baddeley, interference between two tasks is determined by their respective level of learning, we are thus faced with the possible problem of having to distinguish between automatic and controlled processings.

Several demonstrations differentiate automatic processes from non-automatic processes by taking into account many aspects: automatic pro-cessings are quick (Logan, 1988; Neely, 1977; Posner & Snyder, 1975), call for less effort (Logan, 1978, 1979; Schneider & Shiffrin, 1977), and appear to be autonomous and irrepressible, although they can be dominated by conscious processing under particular conditions (Logan, 1980; Zbrodoff & Logan, 1986). According to Bargh (1989, 1996), at least three conditions are necessary for such a domination to occur: the person must (1) be aware of the automatic effect; (2) have the motivation or intention to think or act differently from what comes automatically; and (3) have the attentional capacity to support the flexible, relatively unusual thought or action sequence. Conversely, controlled processings are slow, are used in new situations and in cases where the stimulus and the response are further apart in time, and are consumers of attentional resources (see Camus, 1996; Hampson, 1989; Strayer & Kramer, 1990).

According to Schneider and Fisk (1982), automatic processes directly activate sequences of processing, learned and stored in the long-term memory, and implemented without passing through the working memory. Conversely, controlled processes would require the retention in the working memory of the mnemonic traces of the items involved in the carrying out of the task.

More recently, Logan (1988) has proposed a memory-level view of automaticity, where performance is perceived as automatic when it depends on a direct access to the retrieval in memory of the solution rather than on a kind of algorithmic computation. Based on the idea that every encounter with a stimulus produces a separate representation in memory, leading each episode to be stored and retrieved independently, this approach rests on an episodic conception of memory. By favouring the multiplication of mnesic traces representing similar experiences, the intensive practice of an activity could thus lead to its automation and thus enable, at the expense of an algorithmic mode of processing, the implementation of a direct retrieval in memory. For this reason, performing a task could be described either in terms of a theory of resources or in those of a theory of memory. Whatever the conception of automaticity, according to the more or less automatic or controlled nature of its processing, the carrying out of a task will represent a variable attentional cost.

The Emotion–Attention Relationship.   Tackled by Ellis and Ashbrook (1989), among others, in their studies on memory, this relationship has led to the supposition that a negative emotional state may reduce the amount of attentional resources allotted to the carrying out of a memorizing task. A few studies indeed show that participants in whom a negative emotional state is induced present lower recall performances than participants in a control group (Ellis, Thomas & Rodriguez, 1984). It has also been observed (see Ellis & Ashbrook, 1989) that a negative mood has a harmful effect on activities necessitating cognitively costly encoding operations (difficult or complex operations, for example).

How must we interpret this phenomenon? Ellis and Ashbrook (1989) think that a negative emotional state should increase processings irrelevant to the carrying out of a task, thus reducing the capacity available for the appropriate processings. The capacity would therefore not be reduced, but invested elsewhere. As a matter of fact, Seibert and Ellis (1991) show that not only negative emotional states, but also positive moods generate thoughts irrelevant to the carrying out of the task. They also observe a positive correlation between the participants' thoughts and the memory-recall performance.

Paulhus and Levitt (1987) have examined this phenomenon from another angle. Participants, who were to produce a self-evaluation from word-features with variable valences, were exposed to emotional distractors (words connoted either positively or negatively). Paulhus and Levitt noted a positive self-presentation bias due to the presence of the distractors. For these authors, the salience of the emotionally charged distractors appear to divert the participants' attention away from the processing of the word-features, which were analysed more superficially. Participants would use the more habitual forms of self-description (positive descriptions). An experiment conducted by Paulhus, Graf, and Van Selst (1989) points in the same direction. They show the same type of effect with the manipulation of an attentional load without an emotional implication.

As factors of manipulation of the attention, emotional states thus seem to modify performances by provoking a distraction, as noise would do for example (see Richard, 1980; A.P. Smith, 1991). Such a conception goes back to the late fifties, when Easterbrook (1959) suggested a theory according to which an emotion produces an increase in activation liable to facilitate or, on the contrary, to inhibit performance with respect to the task caracteristics (see Chapter 6 for more details).

As is suggested by these various studies, the existence of a modulation effect of emotions on attentional states gives social insertions a psychological status not to be neglected, in particular, in the study of cognitive performances. *If the retrieval and activation of autobiographical memories are indeed accompanied by the emotional states corresponding to these memories, then the social and emotional history becomes, by way of the use of attentional resources, a source of influence on cognitive performances and activities.*

The allotment of attention to a given cognitive task depends on the interaction between several factors. One such factor is related to the representations of objects and to past experiences encoded in permanent memory. Certain models suggest that the activation of these representations depends on the degree of compatibility between the information linked to the current processing and that stored in permanent memory during previous processings. The confrontation with information similar to other information already processed appears to bring about a comparison with representations previously constructed. This comparison would thus render the representations compatible or not with the new information. In the first case, the latter would not enter the attentional focus. In the second case, individuals would direct their attention towards the information that appears customary to them. Thus, according to Cowan's habituation hypothesis (1988), an incompatibility between representations stored in permanent memory and current information would involve a costly attentional processing. More generally, a spontaneous attentional focusing appears to be the consequence of an incompatibility between new information to be processed and the corresponding activated schema. Participants would then engage in a processing prodigal in attentional resources. In the case of information compatible with, or easily assimilated to, the activated schema, the participants' attention is at first picked up by it, only to turn away again very quickly. Moreover, functional models suggest that the dividing up of attentional resources allotted to a cognitive task may be modulated by the participants' emotional state (Ellis et al., 1984; Kahneman, 1973; Seibert & Ellis, 1991). For example, anxiety increases the participants' level of activation, thus excluding from the field of attentional focus the information relevant to the carrying out of the task. Since negative emotional states in general decrease the amount of attentional resources assigned to a task (Ellis et al., 1984; Hasher & Zacks, 1979; Versace, Monteil & Mailhot, 1993), several authors suggest that the attentional capacity is either orientated towards the self (Cunningham, Steinberg & Grev, 1980; Sedikides, 1992; Wine, 1971), or towards external stimuli (Benoit & Everett, 1993). In this last case, a selective processing of information may be observed. Orientated towards the self, the capacity for attention seems to reduce accordingly the resources assigned to the task.

The allotment of attentional resources would thus depend: (1) on the degree of compatibility between the current situation and the past experiences stored in memory; and (2) on the individuals' emotional state, which may direct the attention towards the self and thus deprive the task of the resources needed for carrying it out.

As we will emphasize in our own studies (see Chapters 3 and 4), attention does seem to play an interface role between individuals and their history and the task and its characteristics. The use of attentional resources appears to be governed by interactions between current and past experiences and by the individuals' emotional state. This state can even express itself as the very consequence of these interactions.

Autobiographical Memory and Cognition.   Up until now, the literature has paid very little attention to the possible links between autobiographical memory and general cognition. Yet certain recent studies (Conway, 1987, 1990a, 1990b) urge us to think that memory and autobiographical memories may play a central role in the representation of conceptual knowledge. The question is to figure out exactly what type of concepts may be associated with autobiographical memories (Conway, 1990a, 1990b). Let us imagine two types of concepts: taxonomical categories (supplies, birds, sports) and categories derived from goals (birthday gifts, camping equipment, things to do by the sea). These two classes of concepts are in some ways different. The first class is mainly involved in the classification of objects. Thus it will be represented in the memory by the properties of a concept, properties that are real in any context, as well as associated in the memory with a decontextualized knowledge. Conversely, the categories deriving from goals would be involved in the instantiation of schemas at the very moment of the achievement of these goals (Barsalou, 1985; Barsalou & Billman, 1989). For example, to pull out the available information from the categories derived from goals—such as whom to invite for a formal evening, gifts to present, etc.—will help to plan the reception in honour of the first Doctor *honoris causa* of a new university. The categories derived from goals would tend to be organized in the memory as specific schemas of knowledge.

Moreover, autobiographical memories seem closely linked, in long-term memory, with the knowledge of events: for instance, going to the movies (action in context) is proved to be an excellent retrieval index (Barsalou, 1988; Reiser et al., 1985). In the same way, an event specifically known to an individual ("when I was living with...") provides a quick access to associated general events (holidays in Italy) and accelerates the retrieval in memory (Conway & Bekerian, 1987).

It is from these distinctions that Conway (1987; 1990a) and Conway and Bekerian (1987) put forward, in a series of experiments, the fact that autobiographical memories are associated in long-term memory with concepts derived from goals. Such a result may explain why apparently comparable individuals may show subtle differences in their representation of the same concept. Finally, the authors state that the categories derived from goals make for an easier retrieval of autobiographical memory than do taxonomical categories. In addition, factors such as the memory's specificity and its date seem partially related to the time needed for memory retrieval. Indeed, more specific memories are retrieved more quickly than less specific ones, and the oldest memories more slowly than the recent ones.

By showing that the concepts derived from goals are closely associated in long-term memory with the encoding of the experiences of events, these studies provide a strong indication to support the idea of a link between elements associated with cognition in general, and elements stemming from a personal experience of social contexts. Of course, we may not infer from this a principle allowing a conceptual elaboration contingent on the individual's experience; but we must admit that certain concepts are understandable as knowledge of events.

If we espouse the various points succinctly presented in this chapter, it is obvious that to take into account studies on autobiographical memory could prove to be fruitful for a social psychology that regards individuals' social history as one of the factors explaining behaviour. The extent of the problems to solve is quite important, however. Close collaborations are therefore required, with cognitive psychology, of course, but also with developmental psychology in order to describe and understand auto-biographical cognition.

Is autobiographical memory one of the relevant levels of explanation for a social psychology of cognition? We cannot ignore the reservations generated by this trend in psychology, which even go as far as taking exception to the use of the term itself. According to this line of reasoning, the studies on autobiographical memory are highly questionable and cannot be generalized. It would thus seem advisable to return to the laboratory to work on episodic memory; all the more so since this concept arouses very little opposition on the academic scene (Banaji & Crowder, 1989).

What can one say in the face of this kind of criticism? Firstly, the studies on episodic memory have neglected the meaning of previous knowledge in the recollection process. Secondly, as it is with autobiographical memory that we are involved, it would seem appropriate to concern oneself with the events related to the personal history, which thus take on a particular meaning for the individual.

Moreover, although equally rigorous, the methods used are not enough to help distinguish between studies in each of these two areas (see on this matter Conway, 1990a; Conway & Bekerian, 1987; Dritschel, Williams, Baddeley & Nimmo-Smith, 1992; Reiser et al., 1985). As noted by Conway (1990a), the main difference is that the researchers in the field of auto-biographical memory start from the premiss of a previous knowledge personally meaningful for the individual to try and understand this very knowledge, whereas researchers in episodic memory tend to minimize its importance. Yet hasn't Tulving himself (1983) underlined the fact that the intervention of personal knowledge and meaning cannot be totally elimi-nated in the data of episodic studies? On the same issue, Conway (1991) replied that it would indeed seem logical to consider episodic memory as a sub-part of the autobiographical memory and to keep the term episodic solely for those studies in which the influence of meanings and of know-ledge is completely controlled. Episodic memory's purpose is to deal with the general, indeed even innate, aspects of memory, whereas auto-biographical memory deals with the aspects acquired in specific fields; both contribute to a better knowledge of human memory. It is in this spirit of scientific division of labour that we must press on with the elaboration of the cognitive bases of social insertion.

SOCIAL BASES

To say that what happens to individuals in their everyday life constitutes the basis of their psychological history is something of a truism. To try to describe and understand how this occurs is a little less obvious. Therefore, after having suggested how and by which mechanisms the individuals' personal history can settle itself in memory and be later retrieved, it is no doubt sensible, as a further step, to return to the origin of the process, that is towards those situations that everyday life offers to or imposes on individuals, thus nourishing their history. To belong to a group, to perform a task in the presence of others, to be evaluated, compared, aren't these all real or potential social insertions liable to awaken psychological states or episodes stored in memory with the contents of knowledge they generate and to which they are connected? In other words, captured in the variety of responses that people are capable of giving concerning comparable situations and objects, wouldn't interindividual, as well as intra-individual, differences be the actualized products of "an implicitation effort" carried out by the individual when relating to things and to others in the social environment? Doubtless we can hold on to this idea, and consider the variety of the inter- and intra-individual responses as an adjustment behaviour, depending on strategies and knowledge based, at least in part, on the nature of these products. In the perspective of this book, environments and social contexts deserved to be approached as fundamental elements of the individual's social and intellectual history.

Even though we have started to outline here a conception of a socially inserted individual, the inadequacy of the drafting still allows for a certain vagueness. Without proceeding directly from the sketch to the final print, the evocation of certain contextual influences should however serve to fill out the composition. In the list of situations most liable to induce social contexts salient for the activities of individuals, social comparison situations are indisputably the most fundamental (see for a review Michinov, 1997; Monteil 1994, 1995). We will now accord them special attention.

Social Comparison

In his *Nicomachean Ethics*, Aristotle has already put forward the idea that self-understanding results from a process of social comparison. Omnipresent in everyday life, social comparison is "an almost inevitable element of social interaction" (Brickman & Bulman, 1977, p. 150).

If Festinger's original theory (1954) was mostly concerned with the effects of social comparison on the individual's appraisal of his or her abilities and opinions, that is no longer the case today. The search for a link between the self and comparison situations and strategies is at the forefront of the scientific scene to the extent that we may even refer to a neo-social comparison theory (Wheeler, 1991). A few authors, such as Pettigrew (1967), suggest that we should include social comparison theory in a more general self-evaluation theory (see also Goethals, 1986), while others think it should be integrated in a theory of self-knowledge.

Self-evaluations, feelings, participants' emotional reactions in comparison situations constitute the new dependent variables in researches conducted within the scope of this new perspective (Brickman & Bulman, 1977; Gastorf & Suls, 1978; Tesser, 1980). By concentrating on the study of the effects of several successive social comparisons, certain studies (Masters, Carlston, & Raye, 1985) even offer the possibility of conceiving the products of comparison as elements whose activation or retrieval in memory might allow us to understand the participant's processing of current situations.

In a series of experiments conducted with children, where participants were to read as quickly as possible words presented on a piece of cardboard, Masters et al. (1985) gave two children a different number of rewards. A first child constantly received an equal number of rewards, whereas a second child was rewarded on an uneven basis, which had the effect of installing between them a variety of comparisons founded on equality and inferiority or superiority. Several sessions were set up in such a way that the participants found themselves in a situation of consistency (identical comparison for two sessions), or of inconsistency (a state of inferiority during the first session, but of superiority during the second session, and conversely). The participants' emotional reaction, assessed by analysing facial expressions at various phases of the experiment, constituted the main dependent variable. Results showed different emotional effects as a function of social comparison experiences. The emotional expression observed reflected the most recent comparison experience only when this experience was positive (recency effect). In the case of negative experiences, it reflected the negative emotional reactions as a whole (cumulation effect). Through the understanding of the consequences of a series of identical or different comparisons, this study thus presents the particularity of coming close to everyday life. It also helps us to see, from the emotional reactions observed, the influence of daily social experiences on the construction of the participant's self-representation. In the same spirit, Morse and Gergen's now classic (1970) study also deserves to be mentioned.

Participants were introduced to a person presenting either socially desirable or undesirable characteristics. Half the participants were led to believe that they were competing with this person for a job. According to the characteristics of the person met, participants, regardless of the nature of the situation (competitive or not) modified their self-evaluation by either increasing or reducing their level of self-esteem. Social comparison thus seems to generate effects on participants' self-perception pointing to the self's great sensitivity to the social context, without however going so far as to promote the idea of a totally chameleon-like person. Moreover, the comparison situations seem selectively to direct the participant's attention towards the relevant information with regards to self-schemas (Markus & Smith, 1981). The self thus appears to play a guiding role in the choice of such-and-such a social comparison strategy (Singer, 1966).

Through the use of a gender schema (masculinity or femininity), Miller (1984) highlights, with regard to comparisons established from performances on a test, different choices of comparison targets in men and women endowed with gender schemas. As these differences are a reflection of dominating cultural norms, one may think that the selection and use of comparative information show the influence of these through the gender schemas that record them. From then on, the probable existence of a relationship between self-schemas and comparison strategies should prompt the study of each in turn. By indicating that social comparison may also be the result of automatic behaviour, Masters and Keil (1987), Kruglanski and Mayseless (1990), and Wheeler and Miyake (1992) offer an interesting perspective, close to the issues related to attentional phenomena developed in this chapter.

Indeed, a number of particularly salient configurations of information appear to activate knowledge engrammed in the permanent memory, thus in turn triggering, through a mechanism similar to "pattern-matching", the activation of a social comparison process (Kruglanski & Mayseless, 1990). In such a conception, when two comparison stimuli are closely linked, the accessibility of one would lead to that of the other. Similarly, elements pertaining to a representation of self and strongly associated to a position occupied by a given dimension (for example success in maths), could become accessible with the activation of that very dimension from the processing of an element belonging to it (a maths test). Investigation into the effects of co-accessibility in social comparison and, more generally, the examination of the conditions of activation of the process, should rapidly become part of current scientific concerns. It is indeed necessary to understand better how social comparison, through the situations inducing it and the consequences reflecting it, provides contents for the self and represents a privileged social configuration for the retrieval of autobiographical elements recorded in self-representations or schemas.

Conway, Difazio and Bonneville (1991) have recently disclosed data that could very well point to a link between autobiographical elements and social comparison strategies. In a first study, men and women were led to experience negative feelings by reading sad stories and listening to depressing music. They were then asked to evaluate the level of sadness of the stories. Finally, they were informed that their evaluations were either identical to or more negative than those produced by other participants (situations of comparison based on interpersonal similarity or difference). Participants were to recollect briefly events from their childhood and their adolescence and describe them verbally. In the comparison conditions, results showed that the participants who had experienced a negative emotional state reported more sad memories than did those in a control group. Moreover, in the interpersonal difference condition, men recollected fewer autobiographical sad events than women. However, men did recollect more sad events than women in the interpersonal similarity condition. According to Conway et al. (1991), men as opposed to women seem to have learned a social rule according to which sadness is a state unsuitable for their sexual category. The comparison situation activating the gender schema, by inferring from the interpersonal difference more than from similarity, appears also to activate the social rule that contains the recollection of autobiographical memories marked with sadness.

By pointing to the priming of cognitions through affects, Bower (1981, 1991) was able to show that, compared to individuals in a bad mood, individuals in a good mood pay more attention to self-enhancing information. This "mood congruence effect" casts new light on the understanding of certain comparison dynamics. Indeed, several studies show that, after experiencing a positive emotion, participants, led to compare themselves to others, do so on the basis of a downward comparison (comparison targets inferior to the self). This particular strategy does not entail a need for self-enhancement, as one might suppose from a more classic interpretation of this phenomenon (Wills, 1981). Rather the participants, by avoiding a comparison with a better-off target that could be detrimental to the self, maintain their positive moods (a congruence effect). As a matter of fact, the same studies show that, following a comparison with better-off targets (upward comparison), participants express a somewhat negative mood. Therefore, even though these results may be disputed (see Collins, 1996), the social context (target superior or inferior to self) and the emotional context (positive or negative mood) in which the comparison takes place are probably not without consequences for the cognitions developed therein.

The social context in which comparisons take place and unfold is quite obviously a major element in the understanding of the consequences that these comparisons are liable to elicit for the participants submitting to them. Thus, Marsh and Parker (1984) noted that good pupils, placed in a high-standard school, may come to construct a negative academic self-concept. However, Marsh (1990) showed that a high level of competence in mathematics correlated with a positive academic self-concept on the maths dimension, but that this correlation did not extend to another branch of knowledge, such as English, for example. Moreover, if the school's high standard in mathematics negatively affected the self-concept, this would only be so in maths, and not in English. These few studies confirm certain results presented in the following two chapters of this book, namely those pointing to differences in cognitive performances as a function of the academic prestige of the disciplines involved.

The Self and the Context of Comparison

The private or public nature of the social comparison is often the only contextual element taken into account in researches studying the choice of a direction of comparison, either upward or downward. In Wheeler, Shaver, Jones, Goethals, Cooper, Robinson, Gruder, and Butzine's (1969) experiment, the public context was defined by the fact that participants expected actually to meet the person of their choice. In this condition, the participants did not seek an upward comparison as much as did those who merely had to point this person out in order to know his or her score on a test. It seems that participants try to avoid the upward comparison so as not to be publicly confronted with a more successful other (see also Brickman & Bullman, 1977). Wilson and Benner (1971) qualify this result with their suggestion that the public nature of a situation of comparison will lead participants to direct their choices of comparison differently according to their own positive or negative self-representation. More precisely, they form the hypothesis that, in the context of a public comparison, participants high in self-esteem will more freely seek an upward comparison than participants low in self-esteem. In considering the role of the self in the choices of comparison, the authors thus substantiate Singer's (1966) idea according to which the self should prove to be a strong determinant of these choices.

In Wilson and Benner's (1971) experiment, participants, chosen on the basis of their high or low self-esteem, were informed that the purpose of the experiment was to determine whether the results on a personality test assessing leadership abilities allowed one to evaluate the actual behaviour of a leader. The predictive validity of the test was supposed to be high, average, or low. The scores on the test assigned the participants an average position within the group. The participants were then led to imagine a group activity liable to encourage the study of a leader's behaviour. Half the participants were then to choose, by ranking them, the people they would like to observe in the case where they are not actively participating in the group activity (public context condition). Results showed that more participants tended to seek an upward comparison in a private context (79%) than in a public one (48%). Differences in behaviour were also observed as a function of the participants' gender: more men tended to seek an upward comparison than women. It also appeared that self-esteem had no effect on the choices of comparison in men in a private context, whereas in a public context men sought an upward comparison when their self-esteem was high rather than low, but only when the predictive validity of the test was high. The context of comparison in which the participants found themselves thus influenced the choices of targets of comparison in the same way as personal factors, such as self-esteem or gender. Moreover, the avoidance of the upward comparison observed in women placed in a context of public comparison led Wilson and Benner (1971) to refer to the force of gender roles in occidental culture and, particularly, women's reluctance to engage in situations of a competitive nature (see also Huguet, 1992; Huguet & Monteil, 1994, 1995). It is also possible that the public context, by rendering the roles particularly salient, led women to attach less importance to leadership than men. Despite the various possibilities of interpretation, not only does this study suggest the taking into account of personal and contextual factors in the choices of comparison targets, but it also indicates a study of the link between the participant and the dimension of comparison, which has always been neglected by researchers.

In an attempt to integrate in the same experimental protocol the main factors influencing the choice of a direction of comparison, R.H. Smith and Insko (1987) tested the hypothesis that the social comparison of abilities depends at the same time on the context of comparison (private or public), on the participants' self-representation (positive or negative), and on performance feedbacks (success or failure). Results did not show any interaction effect between the variables manipulated, but did show three main effects. The first effect revealed that more participants chose a person occupying the highest rank in the group in a context of private comparison than in a public context. The second effect showed that more participants chose a "high-rank person" after a success than following a failure. Finally, the last effect showed that more participants made this choice when their self-esteem was high than when it was low. These results confirm the hypothesis that social comparison also has a defensive function that is not only expressed by seeking a downward comparison but also through the avoidance of the upward comparison (see Dakin & Arrowood, 1981).

Very briefly evoked, these studies nevertheless allow us to observe the lack of any connection found by the researchers between the auto-biographical dimension and the processes and consequences of social comparison. By identifying temporal comparisons (i.e., the same individual comparing himself at two different points in time; see Albert, 1977), the influence of lived experiences on the choice and management of strategies of comparison is certainly underlined; however, the explanatory system called upon is almost exclusively attached to the motivational reference. In need of an intra-individual coherence, the individual appears to develop strategies meant to maintain a diachronically unambiguous self-image (Conway & Ross, 1984). If one conceives temporal comparisons (those very comparisons that may call upon an autobiographical dimension or component for the individual) as being as determining as the others, then the description and understanding of how both types of comparisons interact and are jointly used become central problems for future studies.

Through the questioning of the links between social comparison and self-concept, contemporary research seems to now take into account the temporality related to the self. A few authors (Wheeler, 1991) even go so far as to call for a "neo-social comparison theory". The privileged self-concept component, understood through the measuring of self-esteem, is still too much of an emotional component, and the explanatory framework called upon remains too close to motivation theories; while the use of a comparison process is far from being accepted as a function of the individuals' motives alone (Goethals, Messick & Allison, 1991).

In fact, very few studies endeavour to understand the comparison process *per se* (its cognitive cost, its possible automatization), although some researchers are indeed quite capable of providing approaches allowing for such an understanding (Gilbert, Giesler, & Morris, 1996; Monteil & Michinov, 1996). In the same way, few researchers see in the individual's social history a factor for the implementation of comparison strategies. This temporal aspect, necessarily autobiographical, could however prove to be central in the control of social comparison situations, while also helping to sharpen the intelligibility of their cognitive impact.

Social Comparison Dimensions

Mainly grasped indirectly through the choice or the rejection of targets of comparison (individuals or groups), the strategies described up until now are not the only ones studied. Indeed, the manipulation of certain dimensions of comparison (abilities, personality traits, physical characteristics, activities, for example), allow for other observations. In a few cases, the comparison no longer revolves around the individuals *per se*, but rather insists on their defining characteristics. Thus, in order to tone down the effect of a comparison with a better-off other, it might be feasible either to insist on the differences between the self and the other, to attribute the other's superiority to certain particular dimensions (Brickman & Bulman, 1977), or to create a difference from dimensions not directly related to those included in the comparison situation.

Following the perception of an inferiority, the compensation strategy clearly illustrates the modulation of the comparison by means of the bringing into play of the dimensions stated. Thus Lemaine (1966, 1974) has shown that one may compensate for an objective inferiority by taking refuge in the non-measurable, shifting the comparison towards dimensions irrelevant with respect to the situation. The purpose of such a strategy is to choose dimensions allowing the participant to compensate for a perceived objective inferiority through the creation of a new comparison space. These studies thus underline certain dominating factors in the social comparison process: the psychological distance, or the distance perceived between self and others, the comparison dimensions and the context in which the process is inscribed. Tesser (Tesser & Campbell, 1982; Tesser, 1986, 1988) has attempted to integrate these factors along with one's performances as well as the performances of others. He puts forward the idea that the way people feel is at least in part determined by the contexts in which they find themselves. An important part of that context pertains to the quality of the performance displayed by others. The model developed postulates the maintenance of a positive self-evaluation (the Self-Evaluation Maintenance Model) and suggests a systemic perspective where people compare themselves in such a way as to maintain or increase a positive self-evaluation, personal relationships having a substantial impact on these self-evaluations. Two processes inform the model's dynamic: the reflection process, stemming from the Basking in Reflected Glory [BirGing] model (cf. e.g. Cialdini, Borden, Thorne, Walker, Freeman, & Sloan, 1976), according to which individuals have a tendency to identify with winners in order to increase their own standing through someone else's prestige; and the social comparison process (Festinger, 1954).

In this systemic model, the variables interact, influencing the self-evaluation in two opposite directions according to the process used. For example, if the other's positive performance concerns a dimension important for the self and for its self-definition, the comparison process will be activated, but the self will suffer more from the comparison if the other is a close relation. Or, if the other's performance is high, but on a dimension unimportant to the self, the activation of the comparison process will contribute, in the case where the other is a close relation, to a positive self-evaluation indirectly nurtured, through the reflection process, on the excellence of others.

Constructed Social Comparison

In the absence of comparative possibilities (targets or dimensions) immediately available, the individual could either mentally construct a target or retrieve in memory a pre-existing target (a prototype for example). This way of laying out the problem is apparently closer to the studies on cognition or on social perception than to those on social comparison *per se*. We must not, however, fail to mention this type of studies (see Goethals et al., 1991).

Indeed, it seems that this form of comparison is commonly used. The false consensus effect (see Ross, Greene, & House, 1977) is one among its most famous illustrations. This effect expresses the tendency to think of one's habits, values and behaviours as relatively common and shared by many. For example, drinkers, more than non-drinkers, assume that a vast majority of people heavily consume alcoholic beverages. Obvious in the case of opinions and behaviours, the false consensus effect disappears in the case of abilities, but is highly sensitive to values. Marks (1984) has thus shown that people tend to think that their opinions are widely shared and their abilities unique, underscoring the existence of a false idiosyncrasy effect or a uniqueness bias (J.D. Brown, 1986; Campbell, 1986; Goethals et al., 1991; Marks, 1984; see also our Chapter 7).

Goethals et al.(1991) provide an empirically founded interpretation of these biases. These researchers put forward the idea that, during their socialization, individuals develop a need for enhancement that turns performances, reinforcements and other events into episodes associated to their cognitive, emotional, or behavioural consequences, such as mood and self-esteem. We may thus contemplate the idea here that self-esteem is a particular feature of the autobiographical memory. Therefore, the social configurations facing individuals could provide retrieval indices for previous comparison experiences and, hence, for the strategies with which they are associated.

In short, beyond the dimensions initially devised by Festinger (1954), social comparison is clearly at the core of most of the social insertions offered to or imposed on individuals. Thus it is capable of influencing, perhaps even determining, the subjective evaluation of one's own results or of those of others, hence participating in the psychological description of personal experiences and, by extension, in the positive or negative nature of their mnemonic encoding.

Because of its role, it seems appropriate to consider social comparison as central for the study of the social regulation of cognitions, the more so in that it appears early in the individual's development (cf. e.g. Butler, 1990; Frey & Ruble, 1990; Suls & Sanders, 1982).

CONCLUSION

The construction of an argumentation in favour of a social psychology of cognition through the association of a cognitive structure (the autobiographical memory) and a social configuration (the comparison situations) without proof of their theoretical articulation through highly specific experimental demonstrations may seem pretty flimsy. Without providing real proof, the results reported in the next few chapters give, however, a body of indicators sufficiently convergent to allow such an association to become the conceptual basis for a new way of approaching cognition.

This new approach offers two important contributions. The first one, by taking into account knowledge about the self, introduces the individual's historical dimension as a factor of cognitive regulation. The second contribution transforms the presence of others into a social basis from which to direct the cognitive dynamics.

This dual contribution clearly allows one to open a perspective in which the individual's cognitive properties as well as the characteristics of the tasks to be performed or of the objects to know cannot alone account for the differences observed in intellectual productions. The relationship between this "new social individual" and his or her tasks and objects, as a function of his or her previous social experiences, is at least as important as his or her cognitive apparatus and as the level of complexity of the objects to be processed. This link, however, is no less difficult to apprehend, in so far as it engages an interaction between the current state of the individual involved in the activity and the activation in this same individual of a previous state concerning close or similar activities. The interactions between past and present experiences, apparently triggered, as we shall see, by social comparison situations, seem involved in the process of sharing out and of allotment of attentional resources. This process also has a direct influence on the quality of the cognitive acquisitions and productions.

Therefore the quality of the contribution of a social psychology of cognition may very well lead the way towards a true study of a "situated cognition", without however reducing it to the sole situational dimension, but rather by considering the situation as truly belonging to a self-knowledge capable of generating what we call a "cognitive context of self". This is no doubt what justifies the conceptual association of autobiographical memory and the real or symbolic presence of others with the comparisons generated by that presence.

CHAPTER THREE

The social regulation of academic performances

INTRODUCTORY ELEMENTS

Following the publication of Rosenthal and Jacobson's (1968) study on self-fulfilling prophecies in the classroom, several studies were made of teachers' communication of expectations (Brophy, 1983; Brophy & Good, 1970). These studies showed, for instance, that teachers treated high and low achievers differently by calling less upon the latter than upon the former, by demanding less work from low achievers, and by criticizing them more in the case of an incorrect answer. However, these differences in treatment were not necessarily perceived or always interpreted by the students as proof of discrimination. The possibly pernicious effects of the teachers' expectations on the performances of low achievers did not seem systematic. An understanding of the role of the elements pertaining to the context thus became essential to the analysis of classroom performances (Brophy, 1979).

Futhermore, Levine (1983) noticed that social comparison with high achievers usually generated in low achievers a feeling of inferiority, a lack of motivation and a refusal of competition (see also Chapter 8). This negative impact of comparison, however, seemed debatable. Johnson, Johnson and Scott's studies (1978), for example, suggested that in a joint learning situation, the choice of a partner with whom to perform a task (in this case in mathematics) could follow either an upward or a downward direction of comparison. Thus, before deciding on the negative effects of social comparison, it appeared necessary to take into account the context in which the information concerning the competences and performances of the self and of others was delivered (Marshall & Weinstein, 1984).

Provided that the elements of the context were not neglected, the visibility of the comparative information really did seem to influence academic behaviour. In a study conducted with primary school pupils, Pepitone (1972) showed that the more the tasks performed were similar, the more the social comparison within the groups was noticeable. In addition, Rosenholtz and Wilson (1980) noted that the variety of tasks allowed students to prove themselves in several areas instead of in a single one. Since the less able pupils were still capable of succeeding in at least one area, the effects of social comparison were lessened accordingly. These studies, however, also showed that when the teachers believed in the unidimensional character of academic competences, not to say actually of intelligence, the variety of tasks appeared to be without effect on the importance of the visibility of the comparative information, which could thus remain detrimental to the less gifted pupils.

In the same way, because the differences were visible within a group, the content and the difficulty of the tasks determined different perceptions in high and low achievers. The former interpreted the difficulty of a task as the sign of a positive expectation, and thus scaled up their own expectations and motivations. The latter, interpreting an easy task as a negative judgement on their ability, developed poor expectations and motivations; the more so in that, in both cases, the tasks were publicly presented to the students (Marshall & Weinstein, 1984). The differences in the level of difficulty of the tasks assigned modified the level of expectations, and thus of positive self-evaluation, in high and low achievers. These modifications in the systems of expectations also seemed a function of the pupils' own group and of other surrounding groups. Thus Eder (1983) found either a weakening or a strengthening, at least temporarily, of the self-concept in students who, in some cases, rose to a higher-level group, and, in others, joined a lower-level group.

In the seventies, evaluation and the effects of feedback were studied mostly in order to underline the teachers' differing behaviour towards high and low achievers (Brophy & Good, 1970; Weinstein, 1976). In the beginning of the eighties, new preoccupations appeared, which took into account the standard, the visibility, and the field of evaluation of the competence. By indicating the quality of the performance, the standard of evaluation appeared, as much as the performance itself, as a main source of social comparison and visibility (Marshall & Weinstein, 1984). The visibility of the evaluation seemed to have a varying influence depending on whether it occurred in a group or privately. The motivational strategies thus varied from the one situation to the other and according to the type of reward, the effects of which were studied from the goal structures (Johnson & Johnson, 1974). Indeed, a structure where individual competition and reward are carried out separately does not come under the same level of analysis as does a cooperative structure in which the goals are interdependent. In the same way, a competitive structure in which the individual reward depends exclusively on the outcome of an intragroup competition is not of the same nature as a competition in which the individual reward is attached to the results of an intergroup competition.

Taken as a whole, the studies conducted in these fields all agreed in granting cooperative structures the virtue of lessening the perception of interindividual differences, while competitive structures intensify this same perception (Ames, 1981). Therefore, social comparison, almost always absent from cooperative situations, seemed on the contrary always linked to situations structured by competition (Pepitone 1982; Nicholls, 1984).

Moreover, also during the sixties and seventies, several studies, inspired by Ammons' article (1956), dealt with the effects of the intervention of knowledge of performance. This article followed many studies carried out during the first part of the century suggesting that knowledge of performance, in other words evaluation feedback, enhanced performance (for instance Arp, 1920; Brown, 1932; Gilliland, 1925; Manzer, 1935; Waters, 1933). However, several of these studies presented methodological difficulties, and knowledge of performance was generally not used as such. More often, indeed, only pre-performance expectations were manipulated. In addition, results reflected effects bearing little resemblance to the statement of knowledge of performance as a systematically positive factor. By neglecting results indicating highly variable effects, Ammons' (1956) suggestions suffered from the same inadequacy. Therefore the positive effects of performance feedback continued to be wrongly considered as one of the better-established principles in psychology (Pritchard, Jones, Roth, Stuebing, & Ekeberg, 1988). Consequently, the lack of a feedback effect, or perhaps its negative effect, was interpreted as sampling or measuring bias (see Babab, 1990).

The studies evoked here are to be understood as the backdrop to a time where social psychology was just starting to worry about the study of cognitive development (Doise & Mugny, 1984; see Chapter 7), but was still not preoccupied with the influence of social contexts on academic performances, which we will refer to here as cognitive performances. Thus, before closing, it is important to underline that the reality of a classroom cannot be reduced solely to the taking into account of the aforementioned factors. Individual differences in ability, students' and teachers' expectations and perceptions, the nature and difficulty of the tasks, social comparison and visibility, all these factors are indeed liable to combine and interact, forming a complex social system (Marshall & Weinstein, 1984) most likely to act upon performances.

At the time, the state of thought and the studies on the influence of the environmental factors liable to explain academic success or failure were for the most part descriptive. However, any social psychologist could relate the phenomena identified by the authors of those days to important theoretical and experimental areas of academic social psychology. Social comparison, attribution, motivation, expectation, self-perception, social facilitation (see Chapter 6), etc., functioned as a Gestalt for these results, which often stemmed from classroom observation.

However, in spite of, or more precisely because of, this somewhat holistic acquaintance, the link between these phenomenological rather than experimental data and social psychology's theoretical frameworks is not self-evident. Thus, in our work, we will mostly highlight a body of experimental facts liable to nurture the preliminaries to a theory both local and provisional (Chapter 5), authorizing a more organized follow-up of experimental studies and permitting us to engage in a body of thoughts directed towards the emergence of a social psychology of cognition (Chapters 1 and 2).

As has been stated, the account of our studies will thus follow the chronological order in which they were conducted. In this fashion, the reader will be placed in the same situation as the researcher perusing her or his data and thinking about the best way to interpret them. This is also probably why the reader may feel that there exists a loose relationship between the generic character of the issues at the root of the experimentation and the experimentation itself.

EXPERIMENTAL ILLUSTRATIONS

In order to verify whether certain social variables such as position, status, expectations, or social insertion modes could influence cognitive behaviour, an initial set of experiments in the field was carried out by one of us in the late seventies. Its purpose was to show the differing effects of teachers' ideological affiliations on their evaluation of the competences of the students, depending on whether they were handed out the information that these evaluations would be made public or not. Published much later (Monteil, Bavent, & Lacassagne, 1986), the results showed an influence of a public/private variable not only on the quality of the judgements made, but also on the processing times related to those judgements. These first results allowed one to suppose the existence of a relationship between a social context and an evaluation process. In terms of the explanations put forward to explain academic success or failure, the results suggested moreover that students' intelligence or personalities should not be the only elements taken into account. In this respect, the studies on impression formation and on attribution contributed to confirming the existence of a psychologization process later denounced by Leyens (1983) in his book *Sommes-nous tous des psychologues?*

More importantly, the integration of the existence of others (thus going beyond the intra-individual level alone) in order to understand the relationship between context and cognitive activity encouraged to look into the social conditions in which a task was performed as well as into the possible variations these conditions may bring into being. The main idea here was that, in cases where the relationship to the object or task is modified, it also appears that the social conditions will modify its processing. Therefore all that had to be done was to make use of an academic task to draw together the literature and the various questions about academic contexts.

The first experiment (Monteil, 1988, Study 1) aimed at manipulating the social conditions in which the tasks were performed by using a personal social comparison, such as the one defined by Rijsman (1974, 1983), a comparison based on feedback related to the individual's personal merit. Moreover, the importance of the visibility of the comparative information, underlined in the literature as a factor of intensification of the effects of the salience of the social comparison, led us to opt for the manipulation of the situations either of individuation or of anonymity. We were looking into the possible effects of social comparison as well as trying to pinpoint certain conditions of modulation.

In so far as the results played a key role for what followed, we should describe the conditions of production more precisely.

Academic Status, Social Comparison and Public Individuation Expectation

Sixty-four 14- and 15-year-old boys (32 high achievers and 32 low achievers) participated in this first study. Selected on the basis of their academic results during the past three years,[1] the participants, who did not know each other, were tested in groups of 8, with 4 high achievers and 4 low achievers in each condition (groups were formed by drawing lots).

Before attending a standard biology lesson given by a teacher unknown to the students, it was publicly announced, and in such a way as to allow identification, that one half of the body of students was in Level 1 and the other half in Level 4. This corresponded quite well to these students'

---

[1] While the high achievers' average performance was superior to 15 on a 0 to 20 scale, the low achievers' average was never superior to 8.

normal academic status, because, at that time, the French state education system was testing out "group levels", which involved dividing classes into relatively homogeneous groups with respect to the students' various competences. In this first condition, known as the "social comparison condition", each student's level was disclosed by the teacher, thus differentiating him from all the others. Conversely, the second half was informed that the level was the same for everyone without mentioning, however, what that level was. In this "without social comparison" condition, students were led to assume that the others belonged to the same academic level as themselves.

The lesson was then given in two different situations involving either a high or a low level of social visibility expectation. Students were informed either that each one of them would be given an oral test during the lesson or that none of them would be tested (high expectation vs. low expectation of public individuation). For obvious methodological reasons, no questions whatsoever were asked of any participant regardless of the conditions announced. Because of the size of the groups ($n = 8$), the probability that each student could very well be tested in the high social visibility context was maintained almost throughout the lesson (45 minutes). At the end of the session, the teacher commented that he or she had been short of time to ask questions. Immediately after the lesson, a written test composed of ten questions was given to the students. This second phase also lasted 45 minutes. Each student's test was evaluated by four biology teachers in a different order to minimize evaluation biases. The average of the four marks obtained by each student on the written test constituted the measure of their performance. The question thus was whether, and to what extent, this performance was sensitive to manipulations of the learning conditions. Let's see.

Whatever their public individuation expectation may have been, participants in a situation of non-comparison came up with performances consistent with their usual academic status: high achievers performed very well, whereas low achievers performed poorly. However, quite different results were observed in the comparison situation. While high achievers succeeded much better than low achievers in the individuation condition, both types of students obtained similar performances in the anonymity condition. In a complementary fashion, the high achievers' performance appears better in the individuation condition than in the anonymity condition. The opposite effect was observed in the low achievers (see Fig. 3.1).

The results of this first study seemed to show at least two things: (1) social comparison really did influence academic performances; and (2) this influence was modulated by the expectations of public individuation or social visibility in the classroom.

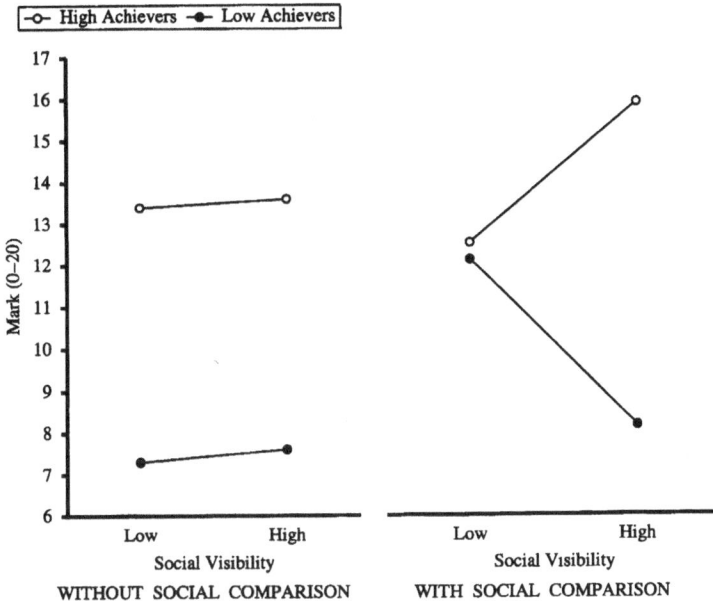

FIG. 3.1   Performance effects as a function of academic level, social comparison, and social visibility in Study 1.

At this stage, all that was needed was to replicate the phenomenon to ensure its strength, which was done in a second experiment (Monteil, 1988, Study 2). Aware that only those situations liable to induce a social comparison were of interest, we could concentrate on testing more specifically the effects of pre-performance failure or success feedback in relation to public individuation expectations (anonymity vs. individuation).

Success Versus Failure Feedback and Public Individuation Expectation

In the second study, conducted only with high achievers, failure and success were attributed randomly from results supposedly obtained on a previous task, constructed on this occasion by the experimenter. Thus the participants were all placed in a situation of personal social comparison. The anonymity and individuation conditions were also introduced.

This extension of Study 1 offered effects almost identical to those previously described. While participants faced with a situation of success did much better in the individuation condition than those facing failure, the opposite effect was observed in the anonymity condition. In the same way, while the performance of students in a situation of success proved to be better in individuation than in anonymity, the opposite effect was observed in students facing failure (see Fig. 3.2).

This interaction between Pre-Performance Feedback (success vs. failure) and Public Individuation Expectation (high vs. low) provided strong confirmation of the effects of the social context on individual cognitive activities. It supported the conclusion that the performance depended neither on the intrinsic capacities of the participants alone (all were

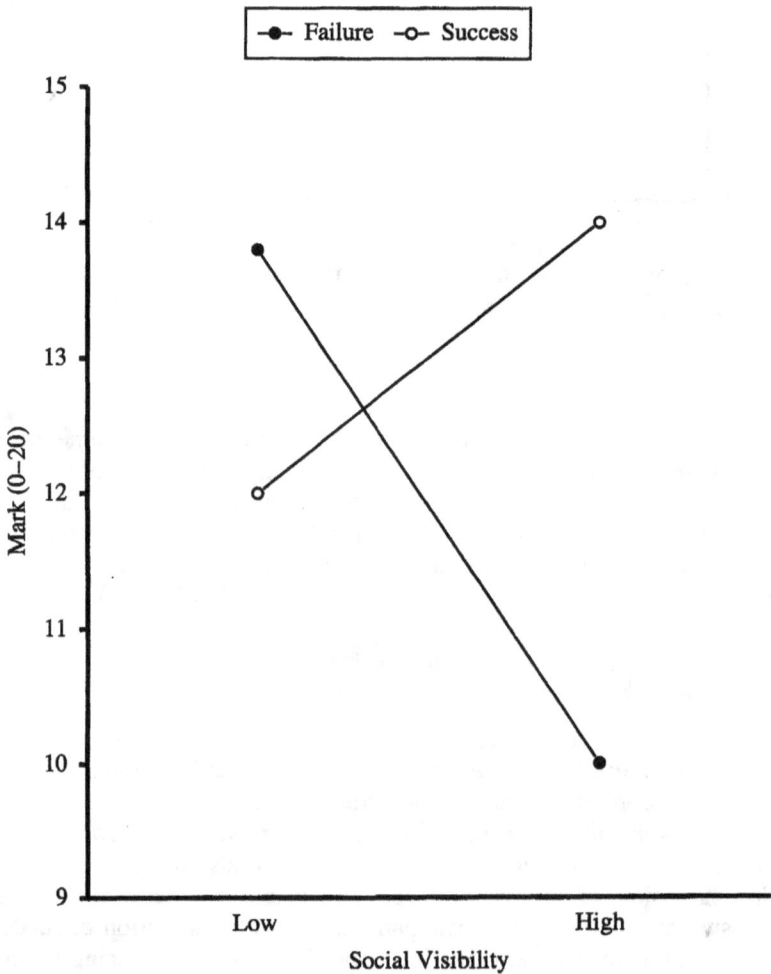

FIG. 3.2   Performance effects as a function of social visibility and pre-performance feedback in Study 2 (high achievers only).

high achievers), nor on the task's own characteristics (it was the same for everyone), but indeed on the context in which the participants worked. This context appeared to determine a particular relationship to the experimental task. The part played by the academic value of the tasks involved in this relationship constituted the object of the next study.

Success Versus Failure Feedback, Public Individuation Expectation and Prestige of the Academic Dimensions of Comparison

In order that readers should grasp the real significance of this third study (Monteil, 1988, Study 3), also conducted with high achievers, we would like to introduce at this point the question of the hierarchy of academic disciplines.

It is well known that the perception of this hierarchy by students and their parents, but also by teachers, causes the school to function as a site of refraction for the social values allotted to each discipline by the dominating cultural practices and discourse (see Chambon, 1990; Mugny & Carugati, 1985).

In reality, there is no scientific argument that would permit us to base a hierarchy of disciplines on a hierarchy of the operations implied by such and such a subject. A plain categorization of types of knowledge according to the specific competences required for the processing of their contents has no empirical basis whatsoever (Simon, 1982). No fundamental difference thus allows us to distinguish the processes called for by the construction and the handling of the various types of knowledge, even when they refer to diversified declarative and procedural knowledge (such as the body of rules and information processed in a mathematical or literary utterance, for instance). The institutional hierarchical organization of disciplines in the school should not be ascribed to the modes of cognitive processing of the various types of knowledge.

For this reason, once the results of the first study had been confirmed, a third experiment, conducted along the same procedural lines as Study 2, aimed to identify more precisely the existence of a social regulation of cognitive performances through the social value of academic disciplines. In this new study, the experimental task thus referred to different disciplines according to the condition: in increasing order of value, Manual and Technical Education (MTE), History and Geography, Biology, and Mathematics (see Chambon, 1990; Huguet & Monteil, 1992). This obviously entailed linking pre-performance feedbacks (success vs. failure) to each of the four disciplines. We were expecting that the interaction effect (Pre-Performance Feedback × Public Individuation Expectation) would emerge, all the more so in that the task concerned academically ranked disciplines.[2] Such was indeed the case.

The size of the interaction effect increased in a linear fashion as a function of the value generally associated with the disciplines involved. While the interaction only accounted for 8.74% of the observed variance in the case of the MTE, the very same interaction explained 22.79% of the variance in History and Geography, 59.90% in Biology and 65.25% in Mathematics! The interaction even disappeared altogether (i.e. was no longer statistically significant) in a little-valued subject (MTE) (see Fig. 3.3).

By way of a modification of the performances, the social stake represented by the hierarchy of academic disciplines thus played, through social comparison, the role of a powerful regulator of cognitive functioning. The social sharing out of school-oriented cognitions thus appeared to strengthen the more or less normative dimension of academic disciplines. As opposed to neutral or less-valued disciplines, the more-valued ones force the student to inscribe his or her conduct in the pattern of expectations generated by the school system. This is why, following these last results, it could be suggested that the modification of an academic behaviour in a given discipline does not depend solely on the students' cognitive characteristics but probably also on the modification of the dominating cultural model at a given time in a given society.

Moreover, it was shown how the perceived value of academic tasks alone can influence performances linked to memory. In the study in question (Huguet, 1992; Monteil & Huguet, 1991), high and low achievers of both sexes, between the ages of 10 and 12, that is to say younger than those chosen in the previous studies, were to learn a complex geometrical figure. Adapted from the "Rey–Osterrieth test", this figure did not hold any particular meaning. It was to be learned in a limited time (50 sec.) and immediately reconstructed from memory by the students (in a graphic form). Each student worked alone, both in the phase of the learning and of the reconstruction of the item. The experimental manipulation consisted in informing participants that the test allowed an adequate measurement of their ability in geometry versus their drawing skills. As opposed to what happened in the previous experiment, only one task was given to the

---

[2] Because the tasks were different according to the experimental conditions, we could not reasonably calculate the interaction between the three variables in the study (Pre-performance feedback; Public individuation expectation; and Type of task). The appearance of an interaction between these three variables would indeed have been difficult to interpret. This explains why the comparison was only on the size of the Pre-performance feedback × Public individuation interaction, according to the type of tasks.

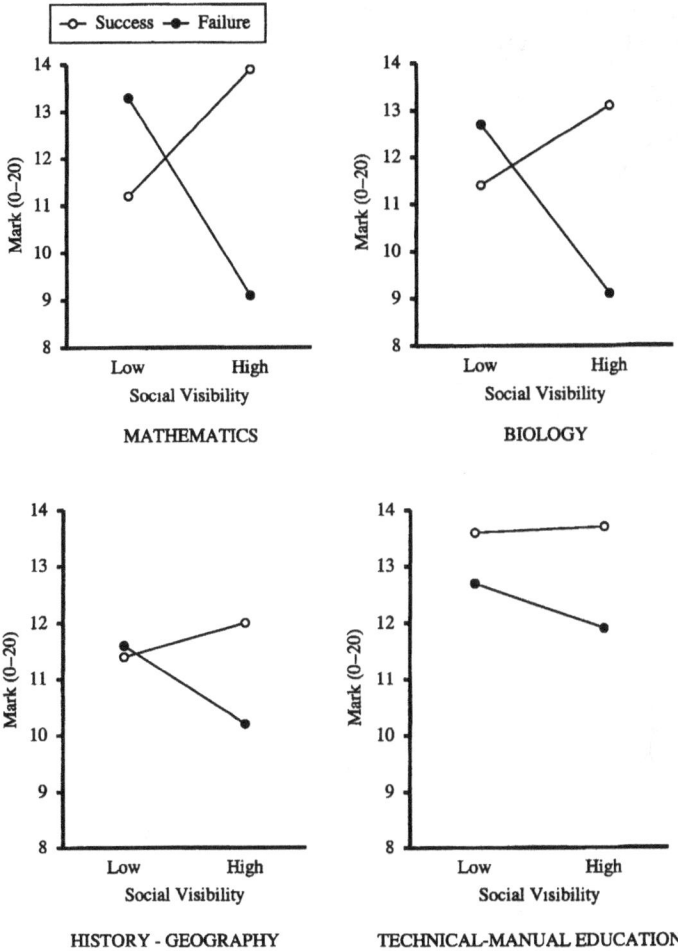

FIG. 3.3 Performance effects as a function of social visibility, pre-performance feedback, and academic context in Study 3 (high achievers only).

participants. The free recall performance was taken as the dependent measure.

Results showed an interaction between the academic level and the context of presentation of the task. In keeping with expectations, the high achievers, in the somewhat prestigious context of geometry, performed much better than in the less-valued context of drawing. The opposite effect was found in low achievers. Moreover, while the mean performance of both groups of participants appeared equivalent in the context of drawing, the high achievers performed much better than the low achievers in the context of geometry (see Fig. 3.4).

The exercise being the same in both contexts, this interaction could only reflect the role of the representations constructed by the participants with regard to the task. As these were young students, the results also testified to the early cognitive integration of the norms and values linked to academic disciplines.

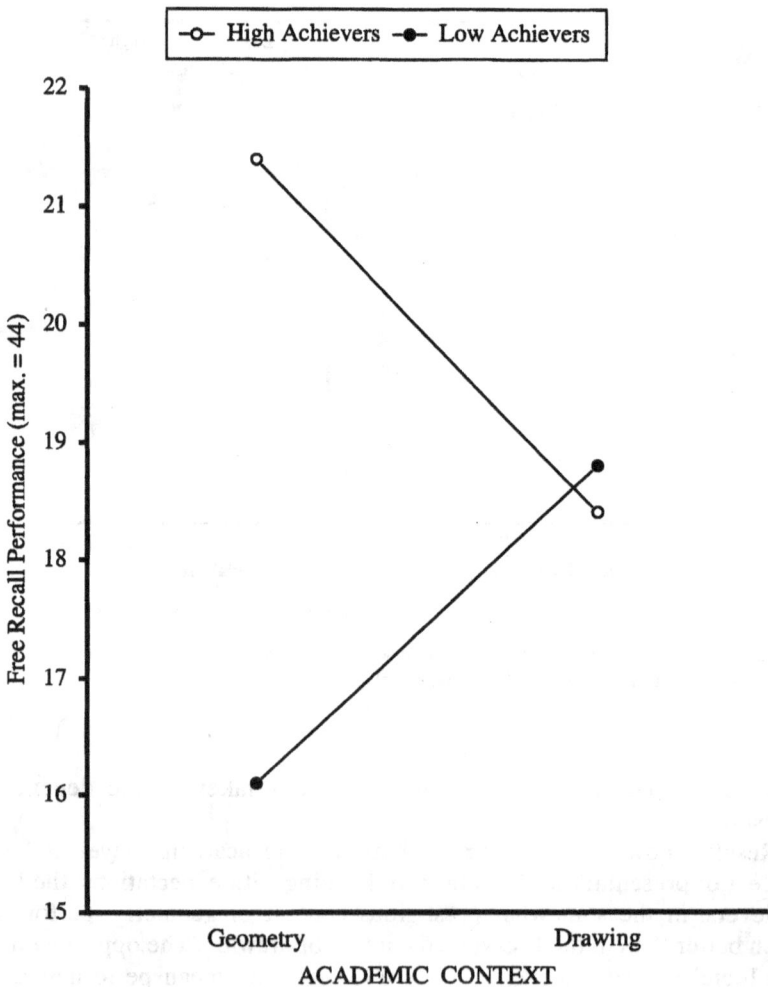

FIG. 3.4   The academic level × academic context interaction on performance.

FIRST APPRAISAL AND NEW QUESTIONS

In a general way, these results as a whole (see also Monteil, 1989, 1992a; Monteil & Huguet, 1993a) thus testified to the importance of comparison situations and to the influence of the academic value of tasks, as well as to that specific insertions, on intellectual performances, without permitting us to identify the causal mechanisms. However, they did indicate that not only was the learning activity modified by the intrinsic properties of the object, but that it could also be changed, at times profoundly, by the social conditions, namely the comparison situations, under which the activity was accomplished.

In fact, not only did these first results induce us to take into account social situations and norms as such; they also encouraged us to include the role of the individual's past experience in the explanation of the effects observed. Indeed, the differences observed in the behaviour of high and poor achievers really did seem to point towards an involvement of the individual's social history. Quite often overlooked in the psychosocial literature, the hypothesis of such an involvement encouraged us to pursue the effort towards a better understanding of the way in which the variables manipulated up to this point could modify cognitive performances and functionings.

CHAPTER FOUR

An "autobiographical attention effect": A second series of studies

The knowledge that individuals have about themselves is no doubt firmly imprinted in their permanent memory. It is probably easily activated and readily accessible, as Tulving's (1983) studies on the influence of episodic contextual information on certain cognitive activities lead us to assume. Hence, because this knowledge is a representational construct, it was feasible to view the social contexts of its development as an integral part of it.

THE ROLE OF ONE'S PAST EXPERIENCES: TWO EXPERIMENTAL ILLUSTRATIONS

This led us to consider that, under certain conditions of activation, cognitive functionings are partially dependent on elements of the representation related to social contexts. It was thus possible to hypothesize that when social insertions (anonymity vs. individuation) were manipulated in participants to whom a value was allotted (success vs. failure)—which value took its full meaning in a social comparison—representations related to the participants' social history were activated.

This could explain why high achievers, when experimentally allotted a success and placed in a situation of anonymity (a state opposed to that which they were generally used to), did not perform as well as in an individuation situation (a more familiar state). It was possible to consider situations conflicting with knowledge that students had about themselves as modifying the social conditions of their relationship to the object (i.e. the task), and at the root of the construction of a new meaning for the social situation.

To tackle this problem of the role of past representations in participants' current cognitive development and productions, it was sufficient, at least initially, to use our experimental procedure with a population of low achievers.

Low Achievement, Success Versus Failure Feedback, and Public Individuation Expectation

In this new experiment (Monteil, 1991, Study 2), adolescents labelled as "low achievers" on the basis of the same criteria as those used in the previous studies, attended a maths lesson that tackled a question new to everyone: Thales' Theorem. The lesson had been standardized beforehand, and was given by a teacher unknown to the students. Following the example of the 1988 Studies 2 and 3 described in the last chapter, a pre-performance feedback from a bogus task was manipulated, randomly distributing success and failure. Students were thus either faced with failure, a somewhat familiar situation, or conversely, with success, a novel situation in view of their past academic experience. The lesson was given either in a condition of anonymity or of individuation. Following the lesson, students were to resolve a series of maths problems making use of the newly acquired information.

The pattern of results was totally opposite to that obtained by high achievers. Indeed, these low-achievers had high performances, unfamiliar to them, in a situation of anonymity and with a success label. Conversely, they performed poorly in a situation of individuation with a success label (see Fig. 4.1).

These students appeared to behave as if, used to academic failure, they had difficulty publicly acknowledging (in a condition of individuation) a positive evaluation (success feedback) that, in a situation of anonymity proved to be rather beneficial to their performances. It is obvious that past psychosocial conditions—negative evaluations and comparisons all through their school years—played a part in the management of the task's current social conditions.

These new results, combined with the previous ones, started to sketch in a more organized picture of the phenomena. Another experiment, conducted in parallel to the four described here, and published later on after replication (Monteil & Castel, 1989), had in the mean time allowed us to show that the effects observed in Study 2 could be modified if, before the allocation of a success or a failure feedback, participants had had a positive experience working in group. This experience appeared to play the role of

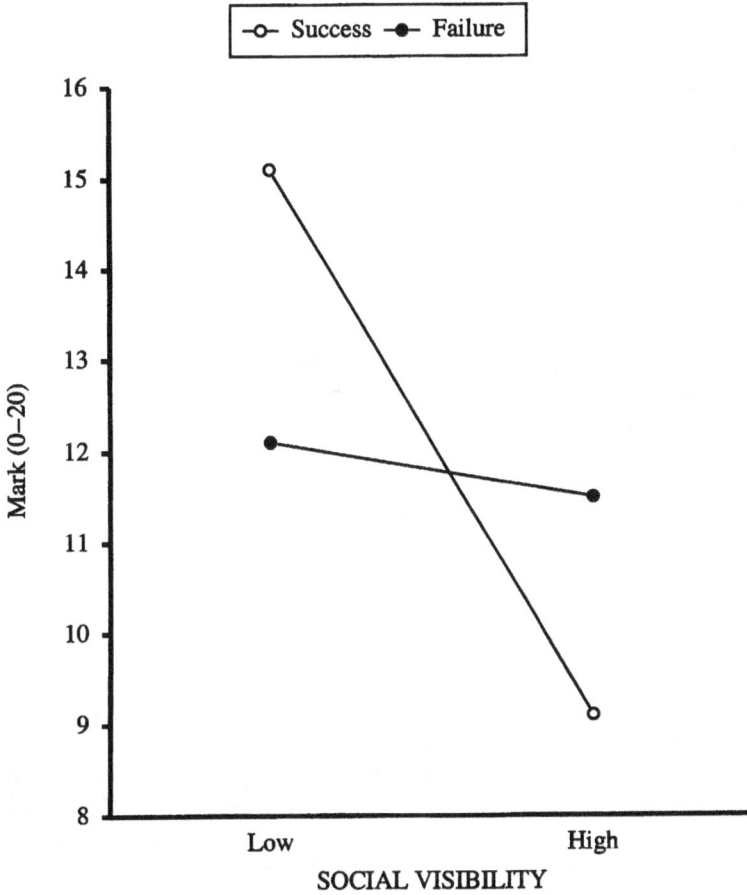

FIG. 4.1   Performance effects as a function of social visibility and pre-performance feedback (low achievers only).

a social resource out of which the participant drew the means to alleviate, among other things, the effects of a negative social visibility.

It thus seemed appropriate, from these bases, to inquire into the existence of mechanisms capable of accounting for, at least in part, the involvement of knowledge and situations related to the self in the production of cognitive performances. It could indeed be imagined that the social situations in which the participants found themselves played the part of a memory-retrieval cue for autobiographical elements possibly related to academic evaluative situations or to social comparisons induced by these situations. Implicitly or explicitly retrieved, these elements, and perhaps their emotional components (which we have begun to show: cf.

e.g. François & Monteil, 1995) appeared liable to play a part in the participant's attentional abilities.

An Integrative Experiment

To test this autobiographical-attentional hypothesis, we used the usual experimental device, to which an incidental task was added (see Monteil, 1993a; Monteil, Brunot, & Huguet, 1996). Let us briefly describe this more integrative experiment conducted with high and low achievers.

As before, students attended a maths lesson on Thales' Theorem in the by now familiar learning conditions (success vs. failure feedback and anonymity vs. individuation condition). On their desks, in the top left corner, a sheet of paper was placed on which a series of small geometrical figures of various shapes and sizes were drawn. The participants' attention was never directed towards this particular feature of their work setting. The experiment ended with problem-solving tasks pertaining to Thales' theorem. Afterwards, participants were required to draw, on a specially provided sheet of paper, the geometrical figures displayed on their desk during the lesson (incidental task).

Results related to the problem-solving task reproduced the phenomena usually observed. When assigned a success feedback, low achievers obtained their lowest score in the individuation condition and their best score in the anonymity condition. Furthermore, regardless of their condition (individuation or anonymity), the failure feedback did not keep performances from being higher than those accomplished in an individuation condition after a success feedback. As for high achievers in the success feedback situation, their results were not as good in the anonymity condition as in the individuation condition. Faced with a failure feedback, their performances were lower in the individuation condition than in the anonymity condition (see Fig. 4.2).

Through the reproduction of results obtained elsewhere (Monteil, 1988, 1991), this last experiment confirmed an already important empirical foundation. However, its main interest did not lie in this confirmation, but rather in the fact of providing, through the recall performances related to the incidental task, the first elements of a mechanism pertaining to the irruption of the personal history into the accomplishment of cognitive performances.

High achievers obtained their best recall score of geometrical shapes in the success feedback condition and in a situation of individuation. In identical conditions, the exact opposite occurred with low achievers, who obtained their poorest score. On the other hand, the low achievers secured, with a failure feedback and either in an anonymity or an individuation situation, their best recall scores. Conversely, it is in these very same

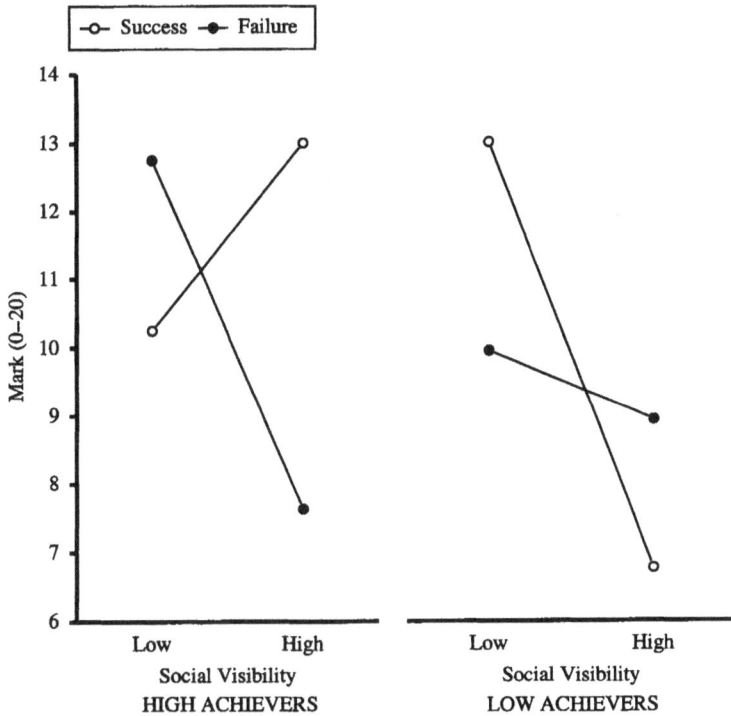

FIG. 4.2   Performance effects (problem-solving task) as a function of social visibility, pre-performance feedback, and academic level.

conditions that high achievers obtained their lowest scores. Finally, the situation in which participants received a success feedback and were placed in an anonymity situation did not yield any difference between high and low achievers (see Fig. 4.3).

TOWARDS AN INTERPRETATION

Although not totally conclusive, this body of results constituted an empirical foundation interesting enough to support the key idea pertaining to these last experiments, namely that the learning conditions in which students are placed can activate knowledge about the self related to past academic situations or events.

Consequently—or at least this was the way we thought we could move forward conceptually (see Monteil, 1993a)—the situations proposed in these experiments and the activated past academic elements, although as yet indeterminate, appeared to the participant as more or less compatible. In other words, these elements seemed to match the representations of self

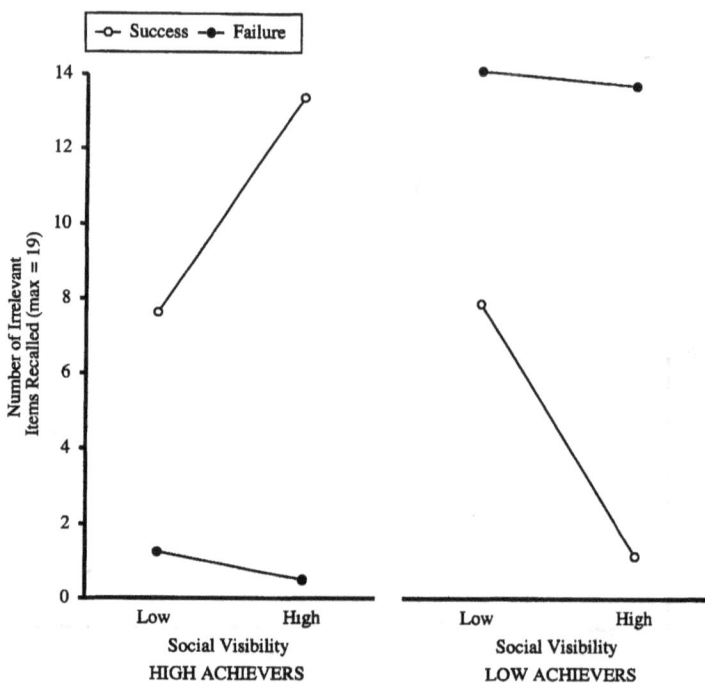

FIG. 4.3   Recall performance effects (irrelevant items) as a function of social visibility, pre-performance feedback, and academic level.

and the autobiographical elements possibly forming these very representations.

According to the degree of compatibility or of incompatibility between the current situation and the activated past states or representations, *the cognitive context of self* thus brought about may represent for the participants a cognitive load and/or emotional characteristics liable to direct the use of their attentional ability. For instance, a strong incompatibility may further the onset of a context salient enough to mobilize an important part of the participants' attentional ability. Consequently, the participants tend to neglect the processing of the problem-solving task—more so when this task requires a lot of attention.

In terms of the progress of the experiment itself, it seems that low achievers receiving a failure feedback, whether in the individuation or in the anonymity condition, were placed in a situation consistent with their academic past. The context, springing from this connection between "past" and "present", was thus so familiar that its processing did not appear cognitively costly. The low achievers were so used to situations of failure and to the ensuing results, in a situation either of anonymity or of individuation, that a good share of their attentional ability was still available. Consequently, the existence of a cognitively "inexpensive" context, coupled with the student's lack of attention due to his/her familiarity with the negative results generally obtained in similar conditions, explained the high recall scores in the incidental task. As it had not been captured by the problem they had to solve, nor had it been mobilized by the context fostered by the experimental conditions, the participants' attentional ability had thus been free to process the geometrical figures present in their immediate physical environment (the desk). Such a result reflected what Carver and Scheier (1981a) have called a "mental disengagement", a phenomenon which can explain the deterioration of performances in the presence of others in cases where the participants evaluate pessimistically their chances of success (cf. e.g. Chapter 6).

High achievers, receiving a success feedback and placed in a position of individuation, also found a highly familiar, and thus cognitively inexpensive, context. As predicted by the cybernetic control theory (Annett, 1969; Carver & Scheier, 1981b), the positive success feedback, coupled with an individuation situation, seems to have set in motion a behavioural self-regulation process capable of explaining the high scores at the problem-solving task. Indeed, as a rule, the triggering of negative feedback loops, supposed to reduce the discrepancy between a behavioural standard activated in memory (under the form of a scheme) and actual behaviour, prompts the participants to focus their attention on the task. If such a process did explain, in part, the high scores at the problem-solving task, it would also have led one to expect low scores at the incidental task, however. In fact, the opposite was observed. The attention still available implied that the strong mobilization of attentional resources linked by definition to the activation and development of a self-regulation process was, in this case, reduced. For the high achievers, the familiar aspect of the situation, combining a positive reinforcement with a situation of individuation, did not require the adjustment of the activated standards of the self to those pertaining to the situation. In other words, the compatibility between, on one hand, autobiographical elements and/or possible self-schemas retrieved in memory by the participants and, on the other hand, the more salient characteristics of the current situation, generated a context demanding very little attention.

All the other situations considered in this experiment presented a more or less strong degree of incompatibility with contents from previous experiences or with the self-schemas that recorded them. As expected, these situations seemed to bring about contexts requiring attentional abilities. Such was the case with low achievers receiving a success feedback and placed in a position of individuation. Individuation, generally linked to failure, coupled here with a success feedback, created, as suggested by the low scores on both tasks, a cognitive and no doubt an emotional context requiring a lot of attention. Likewise, the scores of high achievers receiving a success feedback and placed in a position of anonymity provided a striking illustration of the cognitive load resulting from the relationship between a current situation and autobiographical knowledge linked to past situations. Both in the problem-solving and in the recall tasks, these participants obtained significantly lower scores than in the situation of individuation. Anonymity being highly unusual for these high achievers accustomed to socially visible successes, this new situation disturbed their self-representations, thus contributing to the emergence of a cognitive context of the self that required attention.

This still leaves the problem of understanding how low achievers, receiving a success feedback and placed in a position of anonymity, performed well at the problem-solving task. Indeed, although the low recall scores (that is, the geometrical figures) did indicate the production of a context costly in attention, this cost was not important enough to preclude a positive performance at the explicit task—quite the contrary. Unlike what it produced in high achievers, the position of anonymity put into perspective the context's cognitive load fostered by an unusual success feedback, allowing low achievers to enjoy a positive social comparison. This effect has been substantiated since (Monteil & Michinov, 1998). As suggested by Carver and Scheier's theory (1981a), a positive social comparison appears to implement a self-regulation process directing the participant's attention towards the problem-solving task, thus producing high performances. However, contrary to Carver and Scheier's theoretical presuppositions, the activation and development of this process is carried out *because of rather than despite* the low evaluation aspect due to the situation of anonymity. This last remark may lead us to attach some value to the individual's social history, the importance of which is often wrongly disregarded in the study of problem-solving behaviours.

It seems difficult to ignore the phenomena of the sharing out of attention involved in the modification of the performances observed here. For this reason we have also endeavoured to find out according to what mechanism attention could indeed play a part in the regulation of performances. We showed that students receiving a feedback somewhat inconsistent with their usual performances focused a larger share of their attention on the self than did same-level students who either received no feedback whatsoever, or a feedback quite consistent with their customary performances (Brunot, Monteil, & Huguet, 1995; Brunot, Huguet, & Monteil, 1999).

The tasks' intrinsic characteristics and their context, as well as the motivational or strictly cognitive dynamics prompted by these very characteristics, although inadequate to explain all the variations in performances observed here, led us to take into account the social connection between the participants and the object (i.e. the task) to be processed. This connection, which is mediated by the meaning given to the situation by the participant, seems to be made on the basis of the elements related to the participants' personal history explicitly or implicitly retrieved in memory.

The results observed with low achievers receiving positive feedback in a situation of individuation confirm this interpretation. Indeed, by inducing a strong inconsistency between a given situation and the retrieved autobiographical elements, individuation seems to foster such a cognitively costly context that participants obtain their lowest scores both in the problem-solving task and in the incidental task (Monteil et al., 1996).

Briefly evoked, these last results could no doubt have been interpreted otherwise. We should point out, however, that it was neither implausible nor totally inadequate to view them as illustrative of the involvement of elements related to the self and of autobiographical characteristics in the production of the cognitive performances of individuals. While not altogether laying the foundations for a new theory, these data gave us fresh impetus to develop systematic researches that in turn would offer as many markers as possible to move towards a stronger problematics for the social regulation of cognitive performances. But beyond cognitive social psychology *per se*, our main goal was actually to strive for *a social psychology of cognition* (Monteil, 1993b).

THE SELF AND ITS HISTORY

In order to put forward the idea of the involvement of autobiographical elements concerning the self as constituent parts of an academic success or failure schema, it was necessary to introduce some of the related studies. Following a first exploration into the realm of the self, a doctoral dissertation dealt with the identification of an alleged academic success or failure self-schema (Martinot, 1993).

We already knew that cognitive models of the self were manifold: hierarchical categories (Cantor & Mischel, 1979; Rogers, 1981), prototypes (Kuiper & Derry, 1981), associative networks (Bower & Gilligan, 1979) and schemas (Markus, 1977). Importantly too, the categorizations and evaluations of similar and repeated behaviours in particular fields, either by the self or by others, favour the construction of a perception of the self as a separate being. As a result of this construction, the self-schemas allow the individuals to understand their own social experiences and to process the information stemming from the environment and related to the self. These schemas also have the property of steering or directing attention towards the behaviour most likely to provide information linked to the activated aspects of the self (Markus & Smith, 1981).

Furthermore, when taking into account the extent of autobiographical memory, Kihlström, Cantor, Albright, Chew, Klein, & Niedenthal (1988) suggest that the self is one of the most elaborate and rich structures stored in memory. Finally, studies conducted on the effect, or more precisely the effects, of "reference to the self" provide data that prompt us to think that, in relation to other information, the information coded in reference to the self commands a particular processing. Highly familiar and organizing knowledge, polarized as far as evaluation is concerned, self-schemas thus seem to facilitate the processing and retrieval of events linked to the individual (Greenwald & Banaji, 1989).

Since they stem from a lengthy elaboration and feed on often-encountered similar elements, self-schemas can originate and grow from any personal experience rooted in familiar situations and from repeated practices and reinforcements. It is thus possible to assume that, after several years in school, students possess a self-representation liable to form the subject of a schema.

This line of reasoning led us first to seek academic success and failure schemas and next to identify their involvement in the production of cognitive performances. Using the usual methods of identification schemas (Markus, 1977; cf., for a recent review, Martinot, 1995), we had to struggle in order to grasp even the rudiments of an organized cognitive form of academic failure or success (Martinot, 1993; Martinot & Monteil, 1996). Indeed, whereas participants in a position of success seem to have a cognitive structure displaying the very properties of a self-schema, participants in a position of failure presented less wide-ranging characteristics. In addition, these characteristics appeared less in accordance with the canonical definition, in part because of the positive self-descriptions. Lewicki (1983) interprets this as a support system for a weakened self-esteem; but it may also be understood as the expression of a better access to positive self-attributes. Thus, regardless of the nature of the schema, success or failure, individuals more readily describe themselves with positive rather than negative adjectives and more promptly assess the former as self-descriptive. Although this positive self-descriptive "bias" is naturally consistent with the activation of an academic success schema, this is no longer the case with a failure schema. Socially desirable, positive attributes are also the most cognitively accessible. It is thus normal that, like the high achievers, low achievers, even though to a lesser extent, more rapidly appraise the self-descriptive success items. The differences observed nevertheless permitted us to picture an academic organization of the self differing from the one group to the other. This fact allowed us to measure the impact of success and failure attributions and of social insertions (anonymity and individuation) on the expression of academic failure and success self-schemas' own characteristics. The schemas seemed to us sensitive to social variables (performance feedback and social visibility) related to comparative information (Martinot & Monteil, 1995).

CHAPTER FIVE

Preliminaries to a social comparison feedback theory

The fact of obtaining, through the manipulation of success and failure feedback and of the saliency of the context (anonymity or individuation), significant changes in the cognitive performances of high and low achievers leads us to inquire once more into the functional mechanism(s) at the root of these changes.

A SECOND APPRAISAL

By demonstrating that high and low achievers react differently to identical evaluation and comparison situations, the consistency of our results speaks for the involvement of the personal academic history in the management of the various task-related situations. Because of their saliency, certain situations appear to activate knowledge of an autobiographical type that could very well be stored in such-and-such an aspect of a given academic self-schema. Congruent or not with the participant's current situation, this knowledge would, in all likelihood, create a cognitive context more or less mobilizing the participant's attentional ability, according to the degree of consistency between activated past elements and the current information.

The variations observed in cognitive performances should no longer be perceived as the direct product of the modification of the social context taken as an extrinsic factor. Rather these variations should be understood as the result of a "cognitive contextualization process" that itself results

73

from the activation of participants' personal history under the influence of particular elements defining the social context of performance. It is indeed remarkable to see how cognitive performances, influenced by social contexts and other social insertions, and despite structural, psychological, or physiological constraints specific to each individual, stand completely apart from the competences presumably necessary to their expression. For high and low achievers respectively, failure and success should also be relatively uncommon experiences. But obviously, the results reported or evoked lead us to view failure and success as responses linked to specific social insertions rather than solely to the properties of the task or to individual cognitive competences.

It would be absurd, however, to neglect the importance of any one factor. Even so, by insisting too much on a strict definition of cognitive productions as the result of a direct link between the participant's structural competences and the task's intrinsic properties, the relationship between the participant and the object is disregarded. This relationship does not seem to be a product of the situation *per se*, but is rather the fruit of a personal experience activated in memory on the occasion of a social insertion relevant to the participant. Individuals' social insertion may present a more or less strong salience in relation either to the configuration in which it takes place or to the events or concatenation of events activated in long-term memory (cf. Chapter 2). Therefore, the contents of human memory do not ensue from the processing of the intrinsic properties of objects alone, whether these objects are defined as social objects or not. These contents mostly emerge as products of the relationship between the individual, according to his or her social insertions, and these objects. For this reason, performances, and more generally cognitive functionings, appear to be linked to these contents and to social situations in which individuals use their personal competences and give concrete expression to their life experiences.

Thus, by allowing the expression of a possible sensitivity of cognitive performances to knowledge about the self, the findings reported in Chapters 3 and 4 lead us to believe that the individual or collective social history is indeed a major psychological reality. Consequently, belonging to a group, performing a task in the presence of others, being evaluated, compared, or socially deprived (Monteil, 1992b) are all actual or potential social insertions that evoke psychological states or episodes stored in memory. If we accept this idea, the variety of inter- and intra-individual responses seem to be an expression related to knowledge about the self in part founded on the nature of the elements retrieved thanks to such-and-such a social instance. For this reason, the environment and the context deserve to be treated as fundamental elements of the individual's social and intellectual history. It is therefore important to consider as central the study of the intra-individual cognitive status of the psychological episodes as we have defined them here.

Moreover, as was previously noted in Chapter 2, in the body of situations most likely to foster social contexts salient for the activities of individuals, social comparison situations and the presence of others are on all accounts the most crucial. The presence of others and social comparison are not new to the field of social psychology; but many recent studies seem to have rekindled interest in these issues (cf. Guerin, 1993; Wood, 1996). Facilitated by living in groups, social comparison appears early on in the human being's development (Butler, 1990; Frey & Ruble, 1990; Suls & Sanders, 1982; see Suls & Wills, 1991 for a review). Thus it is important to identify more specifically its particular processes, as we have in fact already started to do (Michinov, 1997; Monteil & Michinov, 1996). Accepted as a "qualifying factor" in psychological episodes, and essential to comparisons (Codol, 1987), the presence of others, real or evoked, necessitates a clear evaluation of its influence on the regulation of cognitive activities and performances (cf. Chapter 6 and 7).

The somewhat syncretic nature of these remarks will certainly not be lost on the reader. It reflects, however, an authentic pursuit of an explanation. A true understanding of cognition and of its products cannot be the concern of approaches that, on principle, rule out any social dimension. A strictly cognitive explanation, probably destined to be "absorbed" by the neurosciences, should not preclude choosing a more integrative level of description.

Consequently, to assemble together, so far as possible, results favourable to one's conception, in order to contribute openly to giving a detailed account of the phenomena before even theorizing about them, amounts to studying the game before playing it. Not as restful as when all the cards are spread out in advance, such an attitude nevertheless lays down a model capable of supplying a minimal theoretical order to the facts observed.

A FEW REMINDERS

Attention

As we saw in Chapter 2, whether attention carries out a function of selection and applies itself to the study of the differential processing of simultaneous internal (memories, knowledge) or external (objects, events) (Johnston & Dark, 1986) sources of information, or fulfils a role in the allocation of available work ability (Kahneman, 1973; Posner & Snyder, 1975), it nevertheless fulfils a prominent function in the processing of information.

The processing of a situation or of a task always entails a variable attentional cost (Fayol & Monteil, 1988). This occurs even in cases where automatic processes are differentiated from non-automatic or controlled processes (Camus, 1988, 1996: Hampson, 1989; Logan, 1988; Neely, 1977; Posner & Snyder, 1975; Strayer & Kramer, 1990), both when automaticity is seen as a memory phenomenon (Logan, 1988), and when performing a task can also be described in terms of a theory of resources rather than a memory theory, according to its more or less automatic or controlled nature.

The Autobiographical Dimension

Individuals' past experiences and the situation in which they find themselves yield part of the contents and organization of long-term memory, and could thus determine how attention is allocated to the various types of information (Nielsen & Sarason, 1981; see Chapter 2). If, as we have noted previously, an organization of a schematic type can regulate this attentional phenomenon, the autobiographical memory can also take the form of a self-schema (cf. e.g. Barclay & Subramanian, 1987; Markus, 1977; Robinson & Swanson, 1989), in other words of cognitive structures that record generic knowledge about the self (stemming from past experiences) and organize and direct the processing of information concerning the self. These structures may include cognitive representations derived from specific events involving the person, as well as more general information stemming from repeated categorizations and from self-evaluations or evaluations by others of one's behaviour. Once formed, this self-organization plays a mediating part in perception, memory, and action (cf. Brewer, 1986; Markus, 1977; Robinson & Swanson, 1990). In this framework, Markus (1977) has suggested that, once established, self-schemas work like selective mechanisms determining the amount of attention paid to the information. By analogy with impression-formation, where individuals seem to direct their attention according to the correspondance between their expectations and the initial behaviours of the target person (cf. Hilton, Klein, & Von Hippel, 1991; White & Carlston, 1983), we could imagine that events different from, rather than matching, a personal recollection call for a more important allocation of attentional resources. Likened to an autobiographical type of structure (cf. Monteil, 1993a), the characteristics of the self would thus become involved in the use and direction of these resources.

Emotion and Autobiographical Dimension

According to the self-discrepancy theory (Higgins, 1987; and cf. Chapter 2), the contradictions, or discrepancies, between what individuals would like to be or deem they ought to be, and their self-concept, appear to represent negative psychological situations linked to certain emotional discomforts. The standards of self are internalized during childhood following the acquisition of knowledge of self and of others. There appears to exist a close link between the individual's standards and emotionally connoted childhood memories (Strauman, 1990).

When activated by elements of a current social situation, some of the knowledge stored in memory appears to induce in the individual the emotional state to which it is linked. Hence, a negative emotional state is liable to reduce the amount of available attentional resources assigned to the carrying out of a given task or to invest it elsewhere (cf. Ellis & Ashbrook, 1989; Ellis, Thomas, & Rodriguez, 1984) thus modifying the learning process (Versace et al., 1993).

Characteristics of the Tasks

As underlined by Kluger and DeNisi (1996), with a very few exceptions, studies about the influence of feedbacks have ignored the theoretical weight of the tasks' characteristics. It has been shown that the feedbacks concerning a motor and a verbal task do not produce the same pattern of results (Annett, 1969) and that the subjective complexity (Ackerman, 1987), as well as the task's academic value (Monteil, 1988; Monteil & Huguet, 1993b), modifies performances. Consequently, these characteristics are obviously possible modulators of the effects of feedback on cognitive performances.

THE SOCIAL COMPARISON FEEDBACK

Up until now we have avoided mentioning the social comparison feedback; it seems appropriate at this point to define it (Monteil, 1997). Stemming strictly from the way in which failure and success have been operationalized in our various experiments, the social comparison feedback may be defined as *the involvement of an outside agent who evaluates the performance, competence or status of an individual, thus placing this individual in a situation of comparison to others.* Seemingly close to Ammons' (1956) concept of the involvement of knowledge of performance, or to other notions where an outside agent acts in a way such as to provide information about certain aspects of one's performance (Kluger and DeNisi, 1996), this definition is nevertheless quite remote from these concepts.

Although it is indeed a matter of knowledge of performance and of competence or status owing to an outside agent, this knowledge presents the added distinctive feature of placing the individual in a situation of social comparison. This difference is of not inconsiderable theoretical importance. Despite conflicting results (Balcazar, Hopkins, & Suarez, 1985; Latham & Locke, 1991), Kluger and DeNisi's (1996) meta-analysis

(131 articles) shows that feedbacks, and specifically performance feedbacks, can be viewed as positive for cognitive productions. The body of studies pointing towards negative as well positive effects, perhaps even towards an absence of feedback effects, fails to take into account the presence of others, either experimentally or theoretically. The studies neglect the possible production of social comparison situations likely to explain the conflicting observations reflected by a thorough analysis of the literature concerning the involvement of feedbacks.

For this reason, strictly reflecting our own results, the preliminaries for a social comparison feedback theory can be founded on six basic propositions or arguments.

Six Basic Propositions

1. Academic production can be regulated by the social comparison generated by the performance, competence, or status feedback.
2. The effect of the comparison can itself be regulated by this comparison's level of social visibility.
3. The effect of the level of social visibility is connected to the individuals' academic history.
4. The interaction between current and past academic experiences can foster an attention-consuming situation.
5. Since attentional resources are limited, attention thus has a hand in the regulation of academic productions.
6. Through modification of the attentional locus, social comparison feedback can thus alter cognitive performances.

These propositions are interdependent, and each one of them is constructed on the basis of the preceding one. Consistent with our data, the perspective adopted here consists in treating the performance behaviour in relation to the social insertions that commit the self and its academic history to its possible emotional content. At the phenomenal level of description chosen here, the mechanism of influence on performances is therefore that of attention referring to individuals' past history activated by the social conditions created by means of the involvement of an outside agent. Consequently, the interaction between the individual's current and past activities can, from a social comparison feedback and according to the more or less important compatibility between these activities, direct the allocation of attentional resources. This direction would be pointed towards such-and-such an activated aspect of the cognitive context of self or towards such-and-such an aspect of the situation put forward. The cognitive performances necessitating a certain amount of attention would therefore be altered.

Consistently with our results (cf. Chapters 3 and 4), it is safe to assume that a strong compatibility of past academic experiences with current conditions of performance entails a cognitive context of the self's costing few attentional resources. The ability to process the elements of the environment remains important in this case. On the other hand, a poor compatibility between past academic experiences and the current conditions of performance induces a more costly cognitive context for the self with respect to attentional resources. The ability to process elements of the situation is reduced—perhaps eliminated all together.

By proposing a schematic representation of the involvement of social comparison feedback, Fig. 5.1 allows us to follow the paths liable to lead to a modification of performances.

In the case of a strong compatibility between past academic experiences and current conditions of performance, the latter necessarily offer a high degree of familiarity and thus entail a somewhat automatic processing. Consequently, the ability to process elements of the situation remains important. For students having a positive academic self-knowledge, their usual expectations lead them to pay attention to the task, especially in high social visibility situations. Faced with a poor social visibility, unusual to them, the cognitive context of the self claims part of the attention available and impairs performance. For students with a negative academic self-knowledge, their expectations, generally linked to failure, lead them as usual to neglect the task, leaving their attention free to process the other elements of their close environment.

In the case of a poor compatibility between past academic experiences and the performance's current conditions, for the students with a positive academic self-knowledge a negative feedback coupled with a strong social visibility creates a cognitive context that is costly in terms of attention. On the other hand, because it fosters conditions of protection with regard to the effects of social comparison feedback, a poor visibility allows these students to pay enough attention to the task to maintain an expected performance.

For students with a negative self-knowledge, a positive social comparison feedback coupled with a strong social visibility, by creating highly unusual open social comparison conditions, leads them to concentrate all their attention on the cognitive context of the self, precluding the taking into account of the task. The opposite happens in the case of poor social visibility, which, by offering protection from having to carry out the task under open social comparison, through positive feedback directs the attention towards the task.

Within this behavioural process, it is likely that some cognitive contexts of the self described here integrate elements of an emotional nature capable of participating in the general organization of the use of attentional

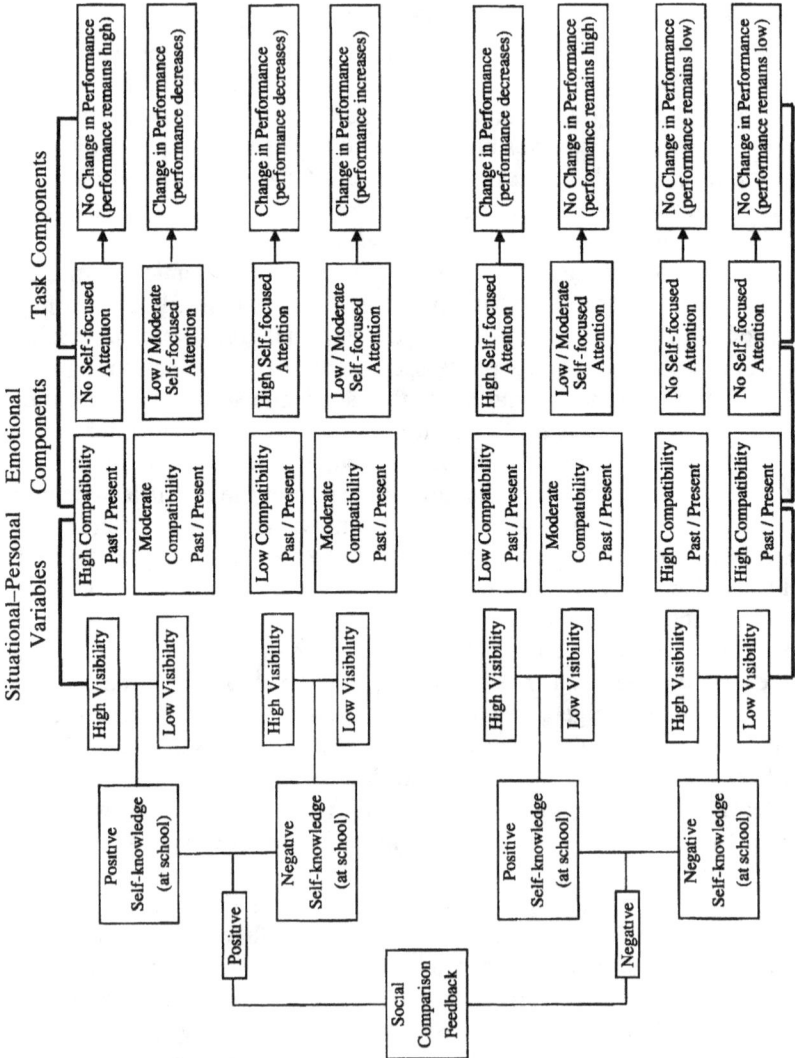

FIG. 5.1   A model for the effects of social comparison feedback on attention and performance.

resources. Moreover, as is shown by several of our results, the characteristics of the tasks probably also contribute to this very organization.

In line with Nielsen and Sarason's (1981) proposition: "One's past experiences and the situation in which one finds oneself would provide the contents and organization of long-term memory and would therefore also help determine how attention is allocated to different kinds of information" (p.946), our outline of the effects of social comparison feedback gives greater importance to a scientific commitment about and over various questions.

Several of these questions have appeared in the course of this chapter; we will mention the most important here, and point to those that are revealing further interesting aspects. What types of autobiographical knowledge (cf. e.g. Conway, 1990a), activated in cases of evaluations and of positive and negative social comparison episodes, are involved in the performance-producing process? As a consequence of the interaction of past and present academic performances, does *the cognitive context of the self* include an emotional component? If such is the case, what is its share of the consumption of attentional resources? Taken as situational variables, are the characteristics of the presence of others and the nature of the social comparison implied by the feedback involved in the cognitive performances, and to what extent?

CONCLUSION

In line with the phenomenal picture offered by the experimental results reported in Chapters 3 and 4, it is likely that cognitive performances are not carried out apart from the personal and thus social history of individuals. We may even presume that performance feedback, characteristics of the tasks, social comparison situations and, to a larger extent, the influence of the presence of others take on their true meaning in the actualization of this personal history dimension, which we will call, using the word in its largest sense, autobiographical.

It is furthermore quite obvious that the attention granted up to now in this book to the cognitive impact of evaluation and social comparison situations, in which the presence of others is a key element, naturally brings to mind the social facilitation and inhibition of performance phenomena. It must indeed be underlined, as we shall see in the next two chapters, that the impact of evaluation and social-comparison-generating contexts on individual productions is not, in itself, something new.

CHAPTER SIX

Social facilitation and inhibition: Markers and thoughts

A long-standing subject of laboratory studies, the influence of the presence of others on individual performance, pertains to one of the oldest fields of research in Experimental Social Psychology. Observed by Norman Triplett as early as the end of the nineteenth century, this at times positive and at times negative influence has recently aroused new interest (cf. Guerin, 1993).

What type of light is the study of this particular influence likely to cast on the results and interpretations expounded in Chapters 3, 4 and 5? Should we assume that social facilitation and inhibition effects pertain to the same mechanisms and processes as those presented so far? Or, on the contrary, are these processes different, while contributing none the less to the social regulation of cognitive functioning? We will get to the heart of these issues in Chapters 6 and 7.

Of course, the different devices used on either side render the processing of these questions rather difficult. These difficulties, however, should not keep us from thoroughly examining the convergences and divergences observed in both cases. The fact of putting them into perspective may be fruitful (Huguet, 1992; Huguet, Monteil, Galvaing, & Charbonnier, in press, c). Fortunately, this does not entail an extremely detailed description of social presence effects, which would be of no use here given the impressive number of studies conducted on these effects since the beginning of the century (several hundreds).

A BRIEF REMINDER

As early as 1898, American researcher Norman Triplett observed that cyclists rode faster when racing together than when racing alone. Likewise, he showed that boys and girls worked faster on a simple motor task (i.e. winding up a fishing reel as quickly as possible) in the presence of co-actors (peers carrying out the same task) than in isolation. Contrary to a widespread idea, however, this "social facilitation effect" does not cover all the results obtained by this author. Triplett, who only mentions the results obtained with 40 of the 225 participants, reports an inhibition effect of co-action in one fourth of his sample and no effects in the last fourth! May it thus be assumed that a differentiated influence of the presence of others is necessarily a trivial phenomenon? Probably not, if one is to believe the studies reported here in Chapters 3 and 4.

The report of a sometimes facilitating and sometimes inhibiting impact of the presence of others on individual performance is widespread in the literature. But still today, this report relies solely on the observation made by Zajonc (1965) that the presence of others facilitates performance on simple or well-learned tasks, and impairs learning or performance on more complex tasks (cf. e.g. Geen, 1991). This observation is the keystone of Zajonc's mostly behaviourist explanation of social facilitation (see below).

But despite its popularity in the scientific community, this explanation comes up against two main difficulties: (i) frequently disputed since 1965, Zajonc's hypothesis can be questioned in favour of more cognitive explanations; and (ii) this hypothesis disposes of the temporal and autobiographical dimension of the social individual. This dimension has sometimes been tackled, but exclusively within the perspective of classical (Cottrell, 1968, 1972) or operant (Guerin, 1993) conditioning. These theoretical reflections, coupled with more epistemological observations, should provide a minimal legitimation to one of this book's propositions: namely that social facilitation should be understood as one of the more basic forms of the social regulation of cognitive functionings.

It is true that the very nature of the tasks (simple vs. complex) is often decisive in the direction, positive or negative, of the variations in performances due to the presence of others. Guerin (1993) reminds us that such an observation was also made, as early as the beginning of the century, by certain education specialists intrigued by the influence of working in groups on students' intellectual productions. Mostly of German origin, these education specialists argued that the presence of peers could either enhance or impair the carrying out of academic tasks, depending on whether the tasks called for creative responses or not. This observation led them to yet another underlying question: Is it best for children to study in class or on their own, in school or at home? Edited and commented on as early as 1910 by W.H. Burnham in the renowned journal *Science*, these studies indeed revealed the superiority of the performances accomplished in class over those carried out at home in the case of activities that did not require much creativity. On the other hand, they pointed to an impairing effect of the class group on certain creative tasks, such as writing a dissertation. To explain this phenomenon, Burnham and his contemporaries ascribed dynamogenic properties to the presence of others and to the evaluations and social comparisons that this presence is liable to generate in the classroom.

In fact, although the studies conducted at the beginning of the century were worthy of interest, they were nevertheless endowed with a poor internal validity. They could not allow any firm conclusion, neither on the strength of the phenomenon of influence in question nor on its explanation. The studies conducted later, in particular Allport's (1920) and Dashiell's (1930), testified both to the reality and to the complexity of the effects of social facilitation and inhibition in human beings, as well as in animals. Their explanatory strength, however, remained inadequate to integrate theoretically many seemingly contradictory results (see Guerin, 1993).

Since 1965, research on social facilitation has soared to such an extent that it has become one of the most important concerns of Experimental Social Psychology. We must recognize, however, that the problem of the very nature of the mechanisms at play and the conditions of their activation and development is as yet unresolved (Huguet, 1993).

ZAJONC (1965): AN INTEGRATIVE HYPOTHESIS AND ITS CONTROVERSIES

According to Zajonc, the mere presence of others increases the level of drive/arousal in the organism thus enhancing the occurrence that the individual will make a dominant response. Therefore, this presence facilitates task performance in the case of a correct dominant response (in simple tasks) and inhibits performance in the opposite case (complex tasks or tasks requiring as yet unmastered skills) (see Fig. 6.1).

To explain how the emission of a dominant response is facilitated in the presence of others, Zajonc used the equation "$E = f(D \times H)$", proposed by Hull (1943)–Spence (1956) in their theory of learning. In this equation, "E" represents the excitatory potential towards a given response, "D" is the organism's level of general drive, and "H" stands for the strength of the S–R connection. According to Hull–Spence, the emission of responses stored in the individual's behavioural repertoire depends on outside stimulations, which may call for either a single response or a set of competing or hierarchical responses. In the first instance, any increment in the value of the energy component of behaviour ("D") will increase both the probability and the speed of emission of the corresponding response. In the second case,

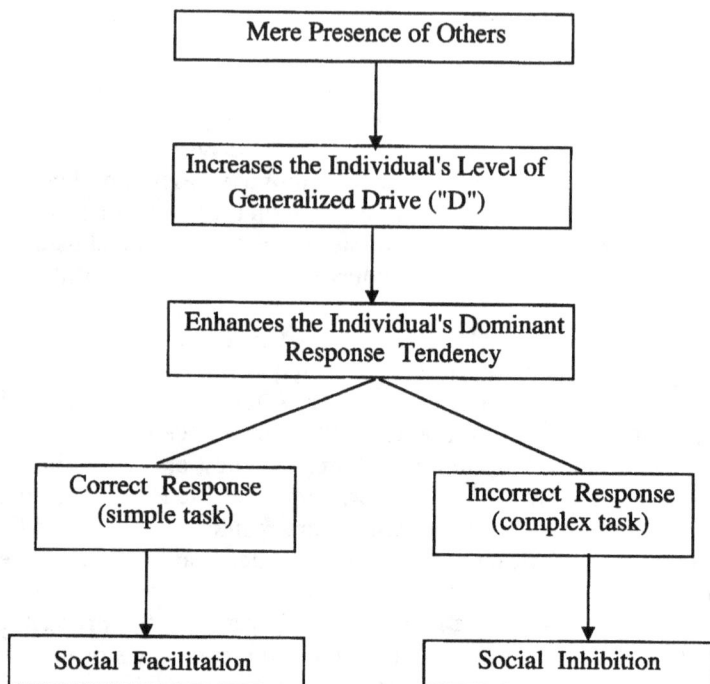

FIG. 6.1   Zajonc's drive hypothesis.

any increase in this value would lead to a corresponding amplification in the hierarchical arrangement of the competing responses. In the long run, this amplification favours the emission of the dominant response and reduces the probability of emission of each surbordinate response.

Based on the detailed examination of 241 studies involving a total of 24,000 participants, Bond and Titus' (1983) meta-analysis provides considerable support for Zajonc's theory. This analysis shows that the presence of others, either as audience(s) or as co-actor(s), generally reduces both the speed and the quality of learning. On the other hand, the performance of simple or well-learned tasks is enhanced in the presence of others. However, there is still much haziness concerning the validity of Zajonc's theory, which has also sometimes met with empirical failure. These issues will be dealt with in the following sections.

Drive and Arousal

To support his solution, Zajonc referred to the results of several studies that allowed one to liken the presence of others to a source of physiological arousal seen as an indicator of the level of drive. Easily accepted, the linear relationship envisaged by the author between arousal and performance might well have been perceived as quite surprising, if only with regard to the Yerkes–Dodson law, according to which this very relation is of more of a quadratic form. Guerin (1993) also denounces Zajonc's restrictive theoretical use of the term "arousal". Indeed, Zajonc only refers to one of the possible uses of the concept, namely the behavioural intensity usage, and forgets to mention how this use can be linked to others (Andrew, 1974 finds six uses).

But more importantly, this aspect of the hypothesis has not been validated since 1965. Bond and Titus (1983) show that only the results in relation to palmar sweat permit the conclusion that there truly is a modification in the physiological state of the participant in the presence of others. Moreover, this modification is only observed when the participant is faced with a complex task. The other measurements taken, of galvanic skin conductance and heart rate, are not very sensitive to the presence of others, without regard to the simple or complex nature of the task. Kushnir (1981) and Moore and Baron (1983) independently arrived at quite similar conclusions.

In fact, such difficulties are not surprising in terms of the behaviourist orientation invoked by Zajonc. As was noted by Glaser (1982), the behavioural arousal imagined by Hull–Spence does not necessarily imply any physiological mechanisms, the concept of drive referring, at least in these two eminent authors' perspective, to a purely mathematical construction. It is true that Spence and his colleagues (Spence, Farber, & McFann, 1956) have sometimes retained, with some success, the level of anxiety, assessed on the basis of verbal reports, as a measure of drive. But later studies using Spence's anxiety scale have invalidated the drive theory (see Glaser, 1982).

Enhancement of Dominant Response?

Serious doubt can also be expressed concerning the very feasibility of a satisfactory operationalization of the concept of "dominant response", which is still widely used in the literature about social facilitation. Guerin (1993), for instance, remarks that the presence of others can facilitate the emission of dominant responses related or not to the task to be carried out, the latter potentially interfering with the behavioural prediction: "If subjects were bored then body shifts and small scratching movements might be dominant responses. If these increased with increased drive then they could interfere with some tasks" (Guerin, 1993, p.37).

More fundamentally, Weiner and Schneider (1971) note the complexity of the constraints inherent in the formulation of any behavioural prediction on the basis of Hull–Spence's theory: "Predictions cannot clearly be derived from the Hull–Spence position without a complete specification of the habit-family hierarchies elicited by the stimuli, without an exact specification of the threshold levels, without an exact knowledge of the oscillating functions, and without a model of conflict resolutions which discloses the probability of each response, given a number of competing responses" (p.259). Most tasks among the ones often retained in the field of social facilitation do not correspond to such criteria. Glaser (1982) has in fact explicitly contested the pseudo-recognition task, used by Zajonc himself in the first test of his hypothesis (Zajonc & Sales, 1966).

In this first study, the authors started by varying the frequency of exposure of participants to words devoid of meaning. The words appearing more frequently were supposed to be more effectively learned than the words appearing less often. The subsequent emissions of the former could be considered as dominant responses and the emissions of the latter as emissions of subordinate responses. Afterwards, the participants were submitted, either alone or in pairs, to a pseudo-recognition test presented as a study of subliminal perception. Stimuli were presented on a tachistoscope at a speed (1/100s) and lighting level such that no perception whatsoever was possible. The participant still had to state the word he/she thought he/she had seen. In line with the authors' predictions, the results obtained show that the participants more often "recognized" the high-frequency words when they answered in the presence of others rather than alone. Put differently, the presence of others favoured the emission of well-learned answers, and thus of the dominant responses in the participant's behavioural repertoire.

One can question, however, the scope of such results with regard to the hypothesis tested. According to Glaser (1982), the pseudo-recognition task, often re-used in the field of social facilitation (Cottrell, Sekerak, Wack, & Rittle 1968; Henchy & Glass, 1968; Paulus & Murdoch, 1971), "is inherently incapable of examining social facilitation effects in the sense of the effects of others in enhancing or debilitating performance, as it involves no more than unavoidably unsuccessful guesswork in response to ambiguous stimuli" (Glaser, 1982, p.267).

According to Kushnir and Duncan (1978), this task allows only one, admittedly pure, measure of response bias as understood in the Signal-Detection Theory, that is, the tendency to respond one way or the other, although there are no real stimuli on the basis of which decisions can be made and the measurements are taken to the detriment of more qualitative aspects of the performance, such as the ability to discriminate between signal and noise (sensitivity). The study conducted by Kushnir and Duncan with a different task (vowel deletion) showed that the presence of others can considerably reduce this discrimination ability without affecting the response bias itself. Interpreted by the authors on the basis of more cognitive perspectives on the concept of arousal (Broadbent, 1971; Kahneman, 1973), this result suggested exploring more systematically the influence of the presence of others in relation to attentional mechanisms.

SOCIAL PRESENCE AND ATTENTION

Conducted in the very field of social facilitation, other studies since 1965 have demonstrated the necessity to call upon more cognitive mechanisms, specially attentional, to explain the effects of the presence of others on performance.

Thus, with five successive invalidations of Zajonc's (1965) hypothesis, Manstead and Semin (1980) suggested tackling these effects on the basis of the distinction made by Shiffrin and Schneider (1977) between controlled and automatic processes. The authors suggest that, as opposed to learning or complex tasks, simple or routine tasks are usually processed automatically. If this type of processing, generally costing few attentional resources, does indeed free some mental space, it can also, on the other hand, keep the performance at a sub-optimal level. By significantly increasing the participants' level of awareness, and hence their control of the activity itself, the presence of others appears to trigger an almost optimal execution of the task, thus facilitating performance. In the case of learning activities, more costly in attentional resources, the taking into account of this presence seems to provoke, on the contrary, a cognitive overload, thus reducing the quality of the attention uniquely available for the accomplishment of the task. This overload almost inevitably leads to a performance decrement on complex tasks, which explains the social inhibition effect so often observed with these tasks. Despite both its relative simplicity and its somewhat integrative nature, this line of research has not been developed as such in the course of the last fifteen years. Only Sanders and Baron's (1975) attentional perspective, compatible with Zajonc's hypothesis, has truly been examined (see below).

In a fashion quite complementary to Manstead and Semin's (1980), Kushnir (1981) suggested that the presence of others might also be processed either in a conscious and quasi-strategic manner or in a way that is perfectly automatic and thus outside the participant's field of consciousness. But again, no empirical demonstration has been put forward to back up this hypothesis. Combined with Manstead and Semin's (1980) proposal, such a hypothesis nevertheless offers certain advantages. Namely, it allows us to contemplate the possibility of a reversal of the effects usually expected from Zajonc's dominant response hypothesis, for instance, the emergence of a social inhibition effect in the case of simple tasks when the presence of others takes on a form unfamiliar to the participant and thus difficult to process automatically. This type of effect, which is at variance with Zajonc's (1965) hypothesis, has in fact been observed in the literature (we will return to this point later on). Kushnir's hypothesis also allows us to understand why the presence of others may, in some cases (see Glaser, 1982), fail to generate an influence on performance. When the presence of others is familiar and can be processed almost automatically, no obvious disturbance in the management of the participant's attentional resources is observed. In this case, the performance stays the same whether carried out in the presence of others or alone.

This notion of a more or less automatic processing in the presence of others, of the evaluations and social comparisons that this presence is liable to induce, is also at the heart of the arguments developed in Chapter 5 regarding some of Monteil's results. The explanation in question, however, throws a more autobiographical light on attentional phenomena in the presence of others. Indeed, it encourages one to view these phenomena as the expression, in part, of a correspondence between current social situations and one or more situations encountered by the participant in the past. In line with this, Glaser (1979, 1982), who found more than 70 failures encountered by Zajonc's (1965) hypothesis between 1966 and 1980, resolved on the necessity of an approach based on the subjective meaning taken on by the presence of others in relation to the self-concepts developed by individuals during their personal history. Thus it is obvious that more cognitive explanations of how social presence affects performance are not lacking in the specialist literature. We must recognize, however, that, no doubt because of the inertia generally associated with the dominant motivational hypothesis, these explanations have been underexploited.

The relevance of a more cognitive explanatory perspective in the field of social facilitation appeared more recently in Hartwick and Nagao's (1990) studies. After clearly underlining the weakness of Zajonc's hypothesis in the field of memory recall, these two authors invalidate it in the domain of memory recognition. From this stems the idea that Zajonc's solution currently seems altogether inadequate in the fields of memory and cognition.

As a matter of fact, a close examination of certain studies on attention and perception conducted before 1965 could have led straight away to an integrative solution of a more cognitive nature with regard to social facilitation and inhibition phenomena.

As early as 1959, Easterbrook showed that, (i) the activation of the more emotional component of drive generally limits the attentional focus to those elements central to the task to the detriment of more peripheral elements; and (ii) this limitation has the effect of either impairing or, on the contrary, facilitating performance according to whether or not it dismisses from the perceptual field the relevant stimuli as opposed to the distractors during the accomplishment of the task. Seldom taken into account in the social facilitation literature (see Baron, 1986; Bruning, Capage, Kosuh, Young, & Young, 1968; Geen, 1976), this attentional explanation has not as yet encouraged any systematic study in this field. As we have already observed, only one hypothesis, formulated by Sanders and Baron (1975), has been thoroughly examined.

According to these authors, social facilitation is a subcategory of a more general phenomenon dealing with the motivational properties of distraction. Sanders and Baron show that the cognitive or motor impact of physical distraction (noise, flash of light, etc.) is indeed often the same as that generated by the presence of others. Fortified by this observation, also made by Pessin as early as 1933, Sanders and Baron conclude that the true antecedents of Zajonc's (1965) motivational mechanism are distraction and the conflict in attention (the problem of sharing out the attentional resources between the task and a distractor) deriving from the presence of others (see Fig. 6.2).

But why does the presence of others generate distraction? Because, according to Sanders and Baron (1975), this presence is capable of satisfying the individual's need for self-evaluation. It is true that, to estimate the relevance of their actions, participants might, for example, try to control an observer's facial expressions. This control can indeed complicate the management of attentional resources for the individual. Along the same line of thought, Baron, Moore, and Sanders (1978) show that distraction can also come from the implementation, by the participant, of social comparison activities with one or more co-actors.

It is, however, difficult from an empirical point of view to decide between Zajonc's perspective and this double hypothesis, both motivational and cognitive. In effect, these two approaches lead to identical predictions: social facilitation in the case of a simple task, social inhibition in the case of a complex task. Moreover, coupled with Zajonc's, Sanders and Baron's (1975) hypothesis is open to the same criticisms (see the discussion in earlier sections). Finally, Baron (1986) admitted that to invoke attentional mechanisms in order to explain social presence effects does not necessarily mean that the drive mechanism should be called upon, as Zajonc (1965) thought.

It is true, however, that the cognitive explanations of social facilitation remain difficult to validate for one simple reason. Following the example of Sanders and Baron (1975), most cognitive perspectives clearly at odds with the drive theory, for example the perspective suggested by Manstead and Semin (1980), lead to the same predictions as Zajonc's. In consequence, even though it is true that several experimental results are somewhat decipherable from a truly cognitive perspective (see Glaser,

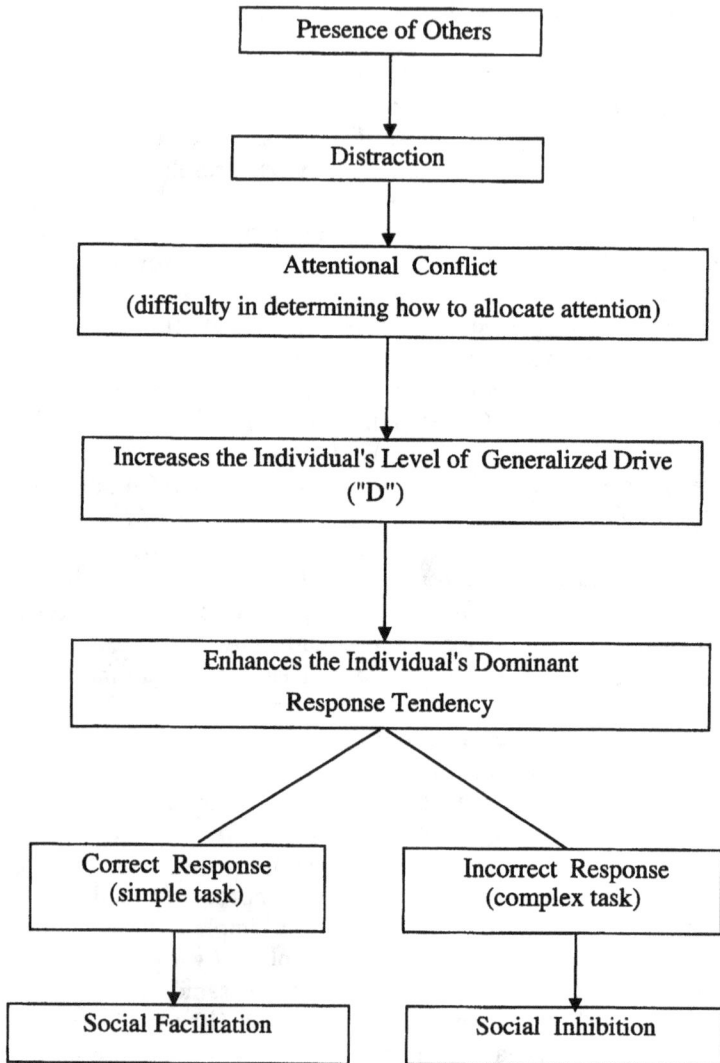

FIG. 6.2 The distraction-conflict hypothesis.

1982), in other words without any reference to drive, these results do not allow us to decide between the theories. To help clarify this important issue, Huguet, Monteil, and Galvaing (1996b) and Huguet, Galvaing, Dumas, & Monteil (in press, b) have recently tested the influence of the presence of others on the Stroop colour–word interference, as suggested by Baron (1986).

Dominant Response Versus Attention:
Recent Experiments

The Stroop (1935) task traditionally consists in showing, for example, the word "red" written in green and asking the participant to state, as quickly as possible, the colour of the ink. The oddity of the stimulus provokes a delay in the vocal RT, sometimes of 250 ms compared to the performance obtained with a more usual (green crosses) or compatible ("red" written in red) sign. This effect is generally explained by the simultaneous activation of the semantic codes "red" and "green", both competing for control of the response (cf. e.g. MacLeod, 1991 for a review). Since the participant's task is not to read but rather to state the colour, the code "red" seems to be selected through an automatic and irrepressible reading process.

From the standpoint of drive theory, the presence of others should favour the emission of the word to the detriment of the colour, thus increasing the colour–word interference. On the other hand, the attentional focus hypothesis mentioned earlier would lead us to expect a decrease of this interference in the presence of others, the attention being concentrated on the colour of the word to the detriment of the word itself.

The results obtained confirm the attentional focus hypothesis. Indeed, the Stroop interference decreased significantly when the task was carried out in the presence of a peer (a confederate) both visually controllable and somewhat attentive (60% of the total time) to what the participant was doing. It blurred slightly in the case where the audience was not visually controllable, and was stable in the case of a more neutral presence (a visually controllable but inattentive peer). Another interesting fact is that it was in the audience conditions favouring a decrease of the interference that the recognition of the Stroop list words was the poorest, the interference and recognition data being positively correlated (the more the interference decreased the more the recognition scores deteriorated). Therefore the decrease of Stroop interference in the critical audience conditions seemed to stem from an early rather than from a late inhibition of the semantic processing of the Stroop list words (the inhibition took place before rather than after the extraction of the meaning of these words). But more importantly, *the reduction of the colour–word interference (the facilitation effect) did not stem from a social facilitation, but rather from a social inhibition of the participants' dominant tendency (word reading)*. In contradiction to Zajonc's hypothesis, this phenomenon, which still has to be unravelled, appears to us as the sign of an elementary form of social regulation of cognitive functionings (Huguet et al., 1998c).

Other recent results (Huguet et al., 1998c, Study 5; Huguet, Galvaing, Michinov, & Monteil, 1997) have also shown a reduction of the Stroop interference in the co-action paradigm. This only occurred, however, when the co-actor (a confederate) was slightly faster than the participant in the accomplishment of the Stroop task, thus placing him/her in a situation of upward social comparison. As in the previous studies conducted in the audience paradigm, the decrease of the Stroop interference was positively correlated to the recognition of the Stroop list words. Furthermore, while replicating such results, we have also discovered (Dumas, Huguet, & Despres, 1998) that the facilitation effect due to upward social comparison in the Stroop task is not mediated by the participants' level of drive/arousal (as measured via heart rate and electrodermal activity during the task). In short, upward social comparison, as well as the fact of being observed by another, facilitated individual performance by inhibiting, at least in part and early on, the processing of the Stroop list words (via an increase in attention focusing on the letter colour cues), even though this automatic processing is extremely hard to repress.[3]

It is now obvious that calling upon the drive mechanism to account for the effects of social facilitation and inhibition is no longer a self-evident step to take. At the heart of Sanders and Baron's (1975) viewpoint, this mechanism can however hardly be invoked to explain most of the results presented in Chapters 3 and 4. Taken as a whole, these results clearly withstand any explanation based either merely on taking into account the task involved or on the relevance of the participant's dominant responses concerning that same task. It should be remembered here that in most of Monteil's studies the tasks assigned to the participants were strictly identical in all conditions. Consequently, from a motivational point of view, no differentiated effect of the situations of evaluation and of social comparison manipulated in these studies should have been observed. But, as we have seen, (1) important variations in performances emerged as a function of the experimental situations; (2) these variations appeared even when the representation of the task rather than its intrinsic characteristics was manipulated; and (3) the effects obtained concerning both performance and attention were overdetermined by the level of compatibility between the present situation and one or more reference situations previously encountered by the participants.

---

[3] Although the Stroop interference may decrease for a number of reasons (an increase in RTs on the neutral items alone, for example), the present effects were exclusively due to faster RTs on the Incongruent Items (Stroop list words) in the critical social conditions with regard to the control conditions. The RTs associated to the neutral items were indeed essentially the same in the experimental and control conditions. As such, our findings are inconsistent with the widespread view reiterated in the Stroop literature that lexical and semantic analyses of single words are uncontrollable in the sense that they cannot be prevented. Instead, it seems that it is possible to prevent the computation of semantics; a point suggesting that mental processes operating outside awareness are not necessarily inevitable (see also Besner, Stolz, & Boutilier, 1997 for a similar argument).

Faced with such observations, how could one not refer to a more cognitive perspective, one that would be more directed towards the social individual's temporal and autobiographical dimension, in order to understand the behavioural effects due to the presence of others? This particular issue also stems from the results obtained in an earlier study (Monteil & Huguet, 1991) conducted with high and low achievers using an experimental design closer to those usually used in the field of social facilitation, as opposed to those called upon in the preceding chapters.[4]

In this experiment, high and low achievers of both genders were to learn a complex figure either in a situation of co-action or alone. Already mentioned in Chapter 3, this figure was presented in the academic fields of art versus geometry. Specifically, the participants were instructed that their task performance would be taken as a means of evaluating their ability either in the field of Art/Drawing or else in the field of Geometry. From bogus scores obtained in a pre-experimental phase, participants were also informed that the members of the experimental categories to which they belonged (identified by the letter "P") were usually, for the same task, better, equal, or worse than those in another category (identified by the letter "O") to which the co-actor belonged (this procedure was taken from Rijsman, 1974, 1983). No mention was made of their academic level.

In the context of Art/Drawing, and only in this context, the poor achievers produced poorer performances when they were the target of a positive categorical comparison than when they worked in the three other conditions, namely control (participant alone), comparison of equality, or comparison detrimental to the self. On the other hand, in the same context the high achievers scored better in the conditions of negative and positive comparison with regard to the two other conditions (control and comparison of equality) (see Fig. 6.3).

As the earlier ones did, these last results encouraged us not to separate participants' personal social histories from the explanation of their performances in the presence of others. The various behaviours observed in high and low achievers are indeed ascribable to the conditions of learning rather than to the nature of learning itself and to the students' abilities. We should bear in mind that since the task was the same in all conditions, participants' behaviours could not be understood with regard to the drive mechanism. Because the task was complex (it was a learning task), the performance should have been inhibited by social presence. Instead, this

---

[4] Although the physical presence of others is one of the key features of the studies described in Chapters 3 and 4, these studies were not designed to test social presence effects as such. Indeed, Monteil's experiments did not include any control, "alone" condition. In contrast, social presence effects were examined in this new study, where such a control condition was included.

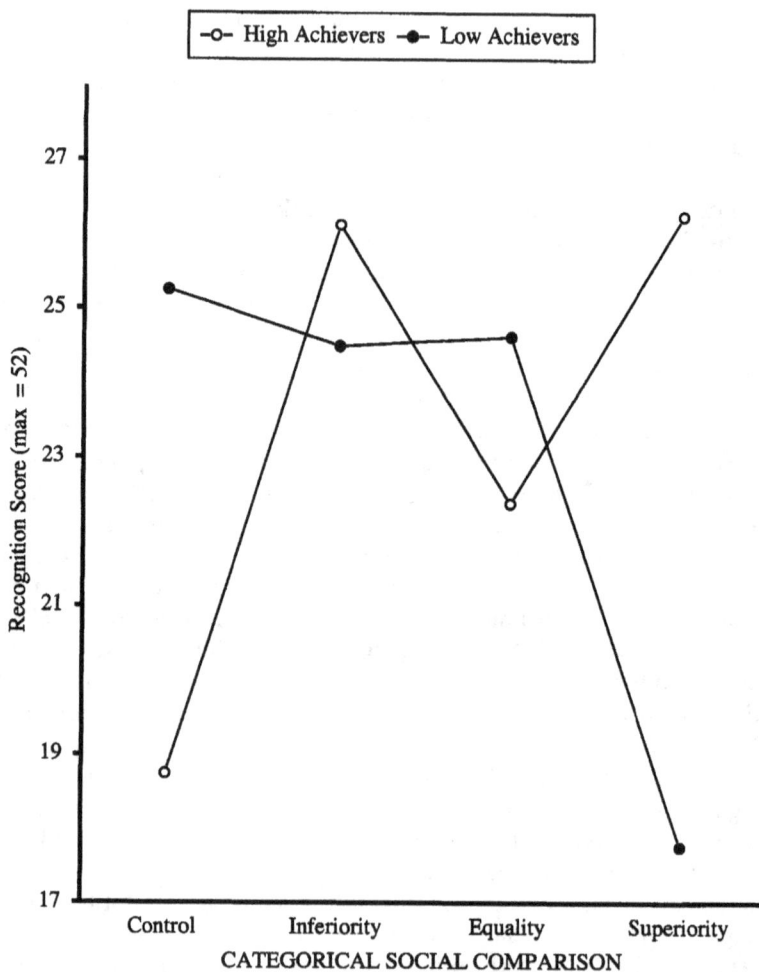

FIG. 6.3   The academic level × categorical comparison interaction on performance (drawing context only).

presence either facilitated or inhibited performance depending on both the social comparison information that was associated with it and participants' academic status. The question of the temporal dimension of the social individual again seemed quite relevant. As before, it seems that the low achievers had difficulty acknowledging a positive evaluation (positive categorical comparison condition) even when faced with a task apparently related to a discipline where evaluative pressure is generally low, relative to most other academic contents, and where they actually perform quite well! Once more, it is obvious (at least for us) that the low achievers' past psychosocial conditions—mostly created by negative evaluations and comparisons all through their school years—played a crucial role in the management of the task's current social conditions.

The question of the temporal dimension of the social individual is in fact the starting-point of an important debate on social facilitation, in keeping with the definition of the conditions necessary and sufficient for the activation of the mechanism proposed by Zajonc. Because of its scope, this debate has no doubt contributed to keeping in the background any extensive questioning concerning the behaviourist basis of the mechanism in question (we will return to this issue later). Exclusively based on theories concerning classical or operant conditioning, the taking into account of social experience in the explanation of social presence effects has provided an important body of experimental results. These results, however, are often just as decipherable, if not more, in terms of a framework of the social regulation of cognitive functioning as in a behaviourist perspective, whether classic or modern.

SOCIAL FACILITATION AND PRIOR EXPERIENCE

Exactly when does the presence of others exert an influence on performance? By implicitly granting to social presence the status of unconditional stimulus, Zajonc (1965) suggested that the "mere presence" of others constitutes a condition both necessary and sufficient to the emergence of social facilitation–inhibition effects. But what exactly is "mere presence"? An abstraction, according to Zajonc (1980), who emphasizes that the mere presence of others cannot be encountered as such in real life. Consequently, "When we speak of *mere* presence in a context of social facilitation, we must mean that performance effects associated with the presence of others can be obtained even though all other factors and processes commonly associated with the presence of others are eliminated" (p. 42). At play in most experimental modern sciences, Zajonc's "molecular" tendency has never been questioned as such. On the other hand, his non-learned drive perspective was disputed from the beginning.

According to Cottrell (1968), individuals seem to learn to respond to the presence of others through a classical conditioning mechanism. Associated in space and time to the positive or negative reinforcements individuals receive on a daily basis, this presence appears to lose its initial nature of neutrality to change into a conditioning stimulus. According to Cottrell, it is thus the anticipation of sanctions, rather than the physical presence of others *per se* ("mere presence"), that constitutes the true antecedent of the motivational mechanism proposed by Zajonc (1965) (see Fig. 6.4).

It naturally follows from this hypothesis that the presence of others, stripped of its evaluative dimension, does not constitute a sufficient con-

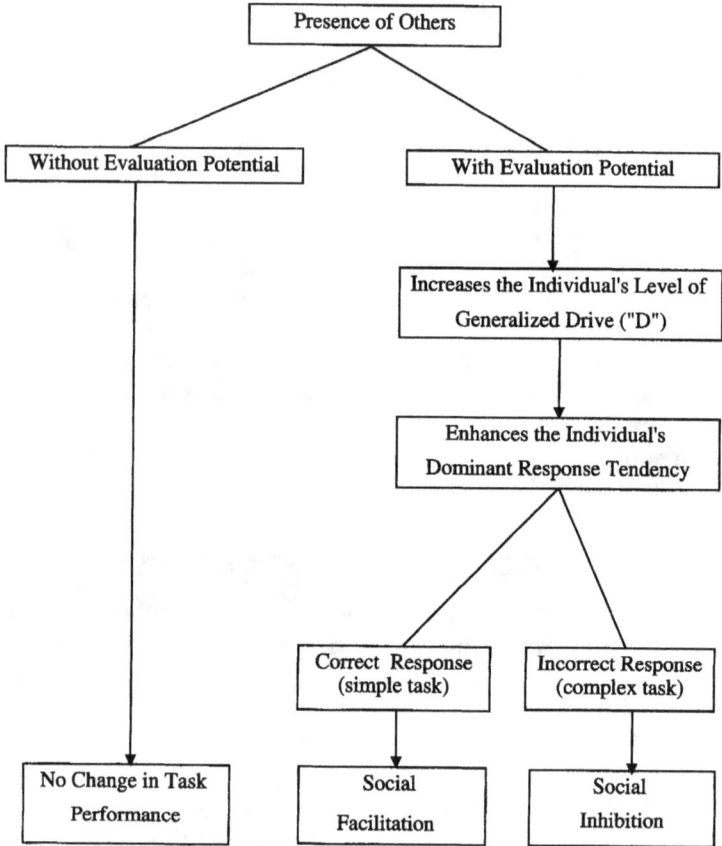

FIG. 6.4   The learned drive hypothesis.

dition for the emergence of social facilitation. It does indeed often emerge (see Geen & Gange, 1977 for a review). However, although several results allow us to attribute to the presence of others the status of a conditioning social stimulus, others, obtained in reference to Cottrell's perspective, encourage us to reject the behaviourist schema that forms its basis.

Geen (1979), for instance, showed that the presence of others, preceded by the allocation of a negative sanction related to the experimental task, was detrimental to the learning of paired words. This same learning, however, was facilitated by the presence of others preceded by a positive sanction (see also Good, 1973). These results were interpreted by the author as a partial confirmation of Weiss and Miller's (1971) hypothesis, which predicts social presence effects only in cases where the participant expects a negative evaluation following the interaction. In fact, such results, replicated by Sanna and Shotland (1990), contrast with the behaviourist orientation that forms the basis of the body of perspectives evoked herein (Zajonc, Cottrell, Weiss and Miller). None of these viewpoints, indeed, allow us to explain the social facilitation effect observed in participants positively sanctioned. In Geen's study (1979), participants were faced with a learning task. According to the perspectives in question, the dominant response facilitated by the presence of others is, in this case, supposed to be incorrect.

Of course, in keeping with Cottrell's model, the influence exerted by the presence of others in Geen's study (1979) does indeed seem to depend on the positive or negative sanction previously given to the participant. The effects observed, however, clearly resist an explanation exclusively founded on the drive mechanism. In the same fashion as the results recorded in Chapters 3 and 4, Geen's (1979) results lead one to think that the meaning attached by the participants to the task and/or to the experimental situation, according to their personal history (limited here to a particular success vs. failure reinforcement), constitutes a significant component of their performance in the presence of others. From a more crucial test of the drive theory, Weiner and Schneider (1971) clearly concluded the necessity of turning to a more cognitive explanation to account for the effects of the expectation of success or failure (and of anxiety) on performance. In another example, Seta and Hassan (1980, Study 2) showed that participants who have a positive history with regard to the task have a better recall performance than participants marked by a failure facing an audience that is supposed to be unaware of the subject's past experience. Both groups of participants nevertheless achieved the same performance when working alone. The participants that had experienced success also produced a better performance when the audience was unaware of their past performance than in the opposite case. Finally, the participants receiving a negative sanction (and only they) produced a poorer performance in the presence of an uninformed audience than when working alone.

In the same way as those described previously, such results hold up against each of the motivational perspectives in question. Thus, only Weiss and Miller's (1971) perspective predicts the last effect noted here. This viewpoint, however, comes up against the absence of effect observed with participants in a situation of failure in the well-informed audience condition. But more importantly, neither Zajonc's hypothesis nor Cottrell's nor Sanders and Baron's allow one to expect that the performance produced in the presence of others will differ depending on whether the participants are placed in a situation of success or of failure (*a fortiori* given that this effect appears only when the audience is unaware of the participant's history in respect of the task).

It is also interesting, at least compared to the contrasting results obtained by Triplett (1898) (cf. the introduction to this chapter), to mention Mash and Hedley's (1975) study. Conducted with young children comparable to those selected by Triplett, this study showed that the influence of social presence on performance for a marble-dropping task can also depend on participants' personal social history. Operationalized by a brief positive versus negative interaction between the audience (an adult) and the participants prior to the carrying out of the task, this history was of a rather minimal nature. This nevertheless proved to be sufficient to modulate the audience effect. Indeed, social facilitation was observed in the case of a positive interaction, and an inhibition effect appeared in the opposite case. This phenomenon is not really decipherable from the point of view of the drive hypothesis. Because the task was simple, the performance should have been facilitated by the presence of an audience in both interaction conditions. Again, such results are more readily explainable once one accepts the idea that the presence of others can generate different meanings, as a function of the participant's history, and that these meanings can direct the individual activity, whether mental or motor.

But even today, the role of the participant's personal social history in social facilitation is examined only in the light of behaviourist models. Thus, supported by Skinner's behaviour analysis, Guerin (1993) suggests breaking with the drive hypothesis to explain social facilitation, otherwise thought to be unexplainable outside of the individual's social experience. The answer championed by the author amounts to considering social facilitation and inhibition as expressions of specific behaviours, selected with the passing of time by the organism according to consequences usually linked, for the individual, to the presence of others in daily life. This answer, seemingly quite close to that previously provided by Cottrell (1968, 1972), is actually very different.

Indeed, according to Guerin (1993), social facilitation is due to the presence of others as discriminative stimuli. As noted by the author: "A discriminative stimulus is not a trigger for behaviour, a learned habit, or an association between a stimulus and a response. If the presence of another person is a discriminative stimulus it means that subjects have previously learned that in the presence of another person the consequences contingent upon the behaviour change in some way" (p.79). In this perspective, no drive assumptions are necessary. "The organism has emitted various behaviours in the past which have been selectively (Skinner, 1981) reinforced with consequences in the presence of other people. This means that in the presence of others, people will respond more frequently with behaviours which have been strengthened in the past when in the presence of others. A drive construct is not needed to derive this" (Guerin, 1993, p.83). The resulting analysis is then "nothing like a Hullian stimulus–

response link, since the context does not trigger the response in any sense, but 'selects' it" (Guerin, 1993, p. 78).

By not resorting to the drive theory, Guerin's perspective (see Huguet, 1993) eludes the criticisms aimed at Zajonc's (1965) hypothesis. For the same reason, it also eludes the difficulties of an empirical nature encountered by this hypothesis. This perspective encourages one to imagine that individuals, because of the specificity of their past social experiences, may respond very differently to the presence of others, regardless of the task. Hence, what can appear as a discriminative stimulus for some individuals may well be non-discriminative for others.

But to resort to Skinner's model of selection of behaviour, as does the author, is not a self-evident move. From an ontogenetic vantage point, this choice leads to an understanding of the influence of the presence of others in terms of an operant conditioning. When it has been dismissed from the explanation of superior mental activities, the use of this framework to understand behavioural variations due to the presence of others may seem surprising. But more importantly, there is nothing from an empirical point of view that compels one to disregard the meanings and representations constructed by the individuals to take into account the involvement of the past in the development of their behaviour in the presence of others.

What exactly is suggested by the results described in Chapters 3 and 4 regarding this important issue? At the very least, they would lead one to credit individuals with the ability to assign a meaning to the reinforcements they receive on a daily basis. It seems that this meaning is greatly para-metrized by the degree of familiarity of these reinforcements and of the situations in which they appear. These results thus lead to the idea of a *piloting of behaviour by the social background of the cognitive system.*

Applied to such results, the explanatory framework suggested by Guerin leads straight away to a dead end. The reason for this is quite simple. This framework would not allow one to predict that behaviours or response systems selected in the past would show the same flexibility as those observed by Monteil. For instance, how should one explain the fact that participants positively or negatively reinforced in the past (poor vs. high achievers) should suddenly alter their performance under the influence of a single historically incongruent reinforcement, even if the situation in which the reinforcement appears is somewhat familiar to them (i.e. the anonymity situation vs. social visibility)? In the same way, how can it be explained that high achievers should alter their performance under the influence of a real but temporary modification of this situation, even if the experimental reinforcement is the same as those encountered in the past? Finally, how should one explain the extreme sensitivity of such effects to the social value of the dimensions of comparison at play (Monteil, 1988, Study 3; Monteil & Huguet, 1991)?

Admittedly, this behavioural flexibility is not easy to decipher in the framework of operant conditioning. This framework integrates the issue of behavioural modification. This modification, however, is supposed to result from a minimal recurrence of new reinforcements in time and space. This is the case neither for the modifications of performances described in Chapters 3 and 4 nor for those generally observed in the very field of social facilitation.

Of course, numerous questions remain for the moment in abeyance. The exact nature of the cognitive mechanisms and of the representations in play still have to be established in Monteil's studies. How and in what form are the participant's past experiences stored in memory? Are these experiences explicitly or implicitly retrieved by the individual? Of course these questions (and there are many more!) are difficult.

But some answers are already in part available. As we have seen, the results recorded in Chapters 3 and 4 allow us to view the personal or categorical comparison episodes generated by the presence of others as an integral part of the mnemonic trace associated with the processing of information, for example the information related to a given lesson in the classroom environment (Monteil, 1993a). In this perspective, the repeated observation of the phenomena referred to earlier led us to conceive social insertions, and more particularly the situations of social comparison, as retrieval cues for autobiographical contents. A part of these contents, as we have seen earlier, may be assigned to self-schemas, as understood by Markus (1977). Such schemas indeed seem to organize and to guide the processing of the information concerning the self and thus to regulate cognitions and individual behaviour (cf. e.g. Markus & Wurf, 1987; Monteil & Martinot, 1991).

Quite different from the models generally invoked to describe social facilitation–inhibition effects, this conception, which integrates the temporal and autobiographical dimension of the social individual, points to the fact that performance may indeed be determined, at least in part, by the level of compatibility between the individual's current social insertions and autobiographical contents retrieved in memory. Results recently obtained (Monteil et al., 1996) seem indeed to indicate that the link thus established between participants' pasts and presents plays a part in the management of their attentional resources at the time of encoding or of recall of target-information.

Of course, it is not a question here of defending the idea that such elements can directly explain social presence effects in all their complexity. But given the current state of the research on social facilitation, one must admit that there is no immediate reason to shun new lines of approach. On the contrary, they are an imperative, in view of the very contradictory conclusions of the meta-analyses on this theme (Bond &

Titus, 1983; Geen & Gange, 1977; Geen, 1991; Glaser, 1982; Guerin, 1986; Sanders, 1981).

More socio-cognitive perspectives, at odds with the drive theory, have already been suggested. Without engaging in any possible comparison with the host of studies (several hundred) referring to the motivational hypotheses, the studies linked to these perspectives may quickly be mentioned.

SOCIAL-COGNITIVE EXPLANATIONS OF SOCIAL FACILITATION

At least two perspectives supported either by Self-Presentation Theory (see Baumeister & Hutton, 1987) or by Self-Regulation Theory (Carver, 1979; Carver & Scheier, 1981a; Scheier & Carver, 1988) explicitly suggest breaking away from the behaviourist orientation in the field of social facilitation.

Self-presentation Theory

Based in part on Goffman's (1959) work, this approach to social facilitation rests on the ideas (i) that the individual develops a positive self-image in daily life and tends to communicate it to others; and (ii) that this self-presentation concern can, in the end, explain the influence of the presence of others on performance. And, as in most motivational perspectives, performance should be facilitated by the presence of others in the case of simple tasks and impaired in the case of complex tasks. Simple tasks will allow the participant to expect a positive outcome to the interaction, thus creating the best conditions for an optimal performance: the participant feels that he/she can save face. By contrast, complex tasks will allow the participant to expect a negative outcome to the interaction, thus creating the worst conditions for an optimal performance.

Several studies do indeed testify to the influence of self-presentation in the development of individual behaviour (Schlenker, 1980) and, more specifically, in the production of motor or cognitive performances in the presence of others (Blank, 1980; Blank, Staff, & Shaver, 1976; Bond, 1982; Dua, 1977; Sanders, 1984). One major conclusion endorsed by these studies agrees with one of our own experimental interpretations: a performance accomplished in the presence of others may depend more on the representation constructed by the participant regarding the task than on the task itself (see Huguet, 1992; Monteil & Huguet, 1991).

Thus the motivational hypotheses have been either partially or completely invalidated by the self-presentation theoreticians. Bond (1982), for instance, shows that a simple task takes more time in the presence of others when it is perceived as difficult and generates a fear of failure. On the other hand, no effect of the presence of others is noticeable when a more difficult task is perceived as easy and generates a success expectation.

Though conducted outside the social facilitation paradigm, other studies have furthered this self-presentation perspective. For example, Baumeister, Hamilton and Tice (1985) showed that performance can either be facilitated or impaired depending on whether the expectation of success is private (personal expectation) or public (expectation communicated by the observer). The presence of an evaluative audience, and moreover of an audience that expects success, can lead to a deterioration of performance, seemingly in order to lower others' expectations. The ensuing new criteria for success will thus appear more conceivable to the individual. The hypothesis of a regulation strategy of social expectations, which posits the implementation of an inferential activity in keeping with self-presentation, has also been successfully tested by Baumgardner and Brownlee (1987; see also Maracek & Mettee, 1972; Mettee, 1971): a pessimistic evaluation of the chances of success seems indeed conducive to modifying social expectations in order to have them match the performances liable to be actually accomplished.

Self-regulation Theory

As opposed to the self-presentation theoreticians (apart from Blank, 1980), Carver (1979) and Carver and Scheier (1981a) explicitly relate the problem of behavioural self-regulation to that of attention, thus allowing them to explain how individuals control and direct their own actions in the presence of others. The organizing principle of this explanatory framework (which corresponds in part with a critical reformulation of Duval and Wicklund's self-awareness theory, 1972) consists in understanding the problem of the links between self-regulation and attention from a cybernetic angle. In this perspective, the control of behaviour appears to be carried out on the basis of a cyclical feedback process, activated by the increase in the level of self-attention, and supposed to unfold according to a sequence of the T.O.T.E. type (i.e. "Test–Operate–Test–Exit"; see Miller, Galanter & Pribam, 1960). The complete unfolding of the process referred to here implies, successively: (a) a phase of evaluation of the actual or anticipated behaviour by comparison with a behavioural standard stored in memory as a schema (phase test); (b) an operating phase allowing either a discrepancy-reduction previously evaluated between the actual or anticipated behaviour and the standard of comparison (negative feedback loop), or a discrepancy enlargement (positive feedback loop) as a function of the positive or negative value allotted by the participant to the salient standard of comparison; (c) a new phase of evaluation, between the actual or anticipated behaviour and the standard of comparison, designed to control the modifications obtained; and (d) a phase of exit in the case of success (a modification evaluated as sufficient would disactivate the control process), and a full reproduction of the cycle in the case of a failure (when the modification is evaluated as insufficient by the individual).

According to Carver and Scheier (1981b), social facilitation may be interpreted as the consequence of a discrepancy-reduction process initiated by the participant between his/her actual or anticipated behaviour and a standard of comparison activated by social presence (see Fig. 6.5). This presence is indeed liable to enhance self-focused attention (Carver & Scheier, 1978). This interpretation of social facilitation has led the authors to consider the presence of others, in keeping with Sanders and Baron (1975), as a facilitating condition among others that are not necessarily of a social nature.

Following this hypothesis, Carver and Scheier (1981b) show that the presence of a mirror as a self-focus enhancer facilitates performance in the same way as would do the presence of an audience (see Innes & Young, 1975; Shaver & Liebling, 1976; Wicklund & Duval, 1971). However, how

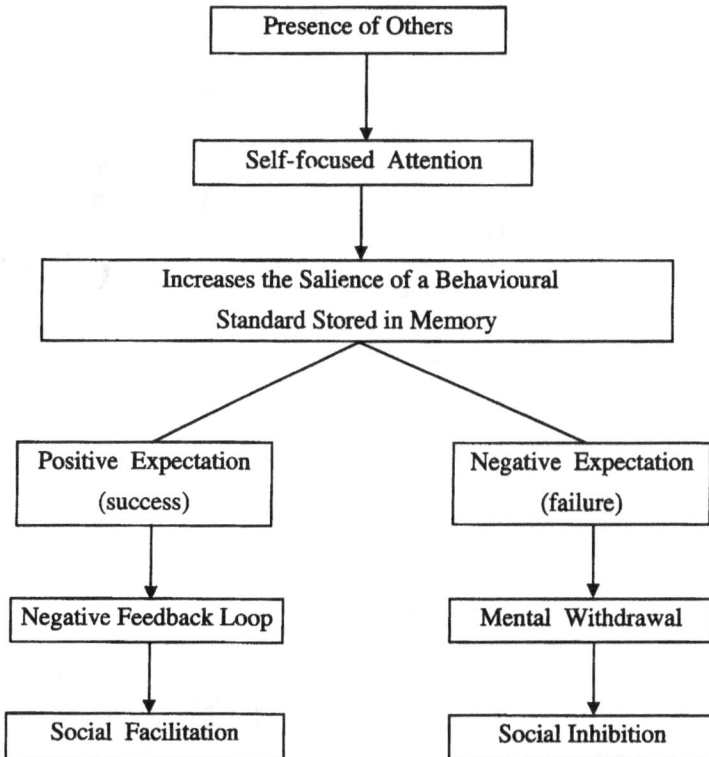

FIG. 6.5   The self-regulation hypothesis.

should one explain the social inhibition effect so often observed in the literature? Carver and Scheier (1981b) suggest that social inhibition is a consequence of a partial implementation of the T.O.T.E. sequence due to a pessimistic estimation of a successful outcome following the test phase (the first phase of the control). Such an estimation may indeed lead the participant to disengage mentally from the task being carried out (see Fig. 6.5).

One may wonder, however, as does Guerin (1993, p.98), about the nature and the contents of these standards and schemas activated by the presence of others in the often rudimentary social facilitation study designs. If indeed self-focused attention (due to the presence of others) can activate certain components of a self-schema, it can no doubt also enhance the salience of self-presentational goals, which is not the same thing. It is not at all obvious that the most basic experimental situations involving the presence of others can generate a behavioural self-regulation of a cybernetic type. Stripped of its social dimension, the situation in play may find no echo in memory.

One can also question both the rationales given for a self-focused attention in the presence of others as well as the role of subjective success or failure expectations in the self-regulation process. Without invalidating Carver and Scheier's (1981a, 1981b) perspective, certain results (Brunot et al., 1995) point towards a more complex self-focus process. More than the presence of others *per se*, it seems that it is the link between the current situation and one or several reference situations previously encountered by the individual that may in the end determine his/her level of self-focus. These results, indeed, show that self-focused attention increases only when the experimental situation is incongruent with the participant's social experience. From this vantage point, it seems that it is the subjective probability of feedback provided to the participant, more than the positive versus negative value of this feedback, that come to affect self-focused attention.

More generally, the socio-cognitive hypotheses, underexploited in the field of social facilitation, remain to be explored. One can only regret that the most recent literature in this field (see Guerin, 1993) does not follow that path, despite the mounting dissatisfaction regularly expressed by numerous authors with regard to the dominant behaviourist explanations of social presence effects.

FURTHER QUESTIONS AND PROSPECTS FOR FUTURE RESEARCH

The very fact that, after a century of research, no definitive conclusion can be drawn concerning the effects of the presence of others on performance testifies to the complexity of the links between social life and human functioning. But has the nature of these links truly been discussed in the specialized literature? Obviously not. This fact in part accounts for the attention, certainly justified although exaggerated, paid since 1965 to the issue of "mere presence" in this literature.

Admittedly, we don't see why social psychologists, as opposed to physicists or to biologists for example, should hold back from collecting more and more fragmentary data, obtained in more and more artificial conditions. In fact, this reductionist trend has already been tried and tested in the very field of social facilitation. We now know that the presence of an observer, even when it is perfectly incidental and stripped of any evaluative dimension, can significantly affect individual performance (cf. e.g. Bond and Titus, 1983; Guerin, 1986, 1993). But because it has tried too hard to obtain socially minimal conditions of experimentation, the literature has progressively abandoned one of the most important questions: that of the epistemological basis of the hypotheses that are confronted with the facts. There is also something disturbing in the fact that the discoveries are deemed to be fragile whenever the experimental conditions are slightly modified, applied to more complex phenomena, or in part moved out of the laboratory.

Epistemological Notes

Let's return briefly to the behaviourist-oriented perspectives invoked in this chapter. By conceiving the presence of others as an unconditional (Zajonc), conditional (Cottrell), or discriminative (Guerin) stimulus, these perspectives lead to the shunning of one level of representation. More generally, they compel us to understand the presence of others as a reactive and derived product of an individual activity exclusively governed by physiological and neurophysiological laws. Obviously, this epistemological stance forces us to endow the social context with an objective reality, with almost universal properties, and with the status of an outside source of variations.

By testifying both to the dependency of individual behaviour on the subject's personal social history, and to the necessity of calling upon a more cognitive perspective to explain the performances produced in the presence of others, certain results (namely those in Chapters 3 and 4 and a few more referred to in this chapter) contribute towards the enhancement of a radically different epistemological stance. According to this position, the social context only exists through the involvement of cognitive structures of contextualization, such as those related to the individuals' autobiographical memory (Chapter 2).

But let us make no mistake. The choice of a more socio-cognitive direction does not straight away guarantee a radical opposition to the conceptions put forward by the behaviourists. Unlike the latter, the self-presentation and self-regulation theorists are less concerned with the properties of the stimuli with which the participant is faced than with the activities and cognitive processes assumed to be activated or triggered by the presence of others. Does this difference, admittedly quite important, give a new epistemological status to the presence of others compared to the one in force in the dominant orientation? We have to admit that it doesn't. In the perspectives in question, the presence of others appears, once again, as an external source of variations. And this very source is this time deemed liable to facilitate or, on the contrary, to inhibit activities and processes neglected by behaviourists.

In conclusion, although this chapter provides no new theory concerning social facilitation, it nevertheless suggests a departure from the underlying epistemology endemic in this field of study. Given the voluntarily selective empirical elements called upon in this chapter, it seems only fair to us to try to integrate the phenomenon in question in the framework of a more general problematics of the study of the social regulation of cognitive functionings (cf. Chapters 1 and 2). This integration suggests new hypotheses hitherto ignored in the literature on social facilitation.

Future Research

The most basic hypothesis amounts to viewing the presence of others as a retrieval cue for self-related information and for autobiographical contents (as in Chapters 2 and 5). From this vantage point, one can study, for example, the order and speed of appearance of categories of autobiographical knowledge in the presence of others. Coupled with the reductionist approach traditionally used in the field of social facilitation, this study should permit us to evaluate the efficacy of social presence as a retrieval cue for information about the self at the various levels of sociality of the experimental situations. In a nutshell, these levels may vary from the presence of a non-evaluative audience to the presence of co-actors set against an explicit social comparison backdrop. From this same perspective, one can also explore the influence of the presence of others on the processing and retrieval of self-related information in self-schema paradigms. Previously suggested by Markus and Kitayama (1991), this study could in particular allow one to evaluate the mediating status of certain schemas in social facilitation–inhibition phenomena. This is an important question about which we are obviously quite ignorant for the time being.

More generally, it may no doubt seem quite unwise to limit the study of social facilitation to target-performances, whether motor or cognitive, as has been the case since 1965. Indeed, as was suggested by Guerin (1993), the simultaneous identification of behaviours without any direct link with the task to be carried out can enrich the interpretation of the effects obtained by the performance. We are well aware that this type of identification should help to put to the test the involvement cognitive structures of contextualization in the development of individual behaviour in the presence of others.

A second complementary hypothesis consists in considering that the involvement of such structures may affect the participant's attention ability during the carrying out of the experimental task. By testifying to the weight of the individual's social history in the management of his/her attentional resources in evaluation and social comparison situations, certain studies described in Chapters 3 and 4 prove the validity of this hypothesis. Its importance lies firstly in the fact that attentional mechanisms are never explicitly studied with regard to autobiographical contents in the field of social facilitation. On the other hand, the importance of such a study lies in the necessity of showing how the involvement of self-related knowledge affects performance in presence of others. In keeping with both this second hypothesis and, more specifically, with Markus and Kitayama's (1991) proposition, the literature on self-schemas suggests that, once formed, such schemas appear to play a mediating role in perception, attention, memory, and action (e.g. Brewer, 1986; Markus, 1977; Robinson & Swanson, 1990).

The impact of social presence on attention and performance should also be studied from the point of view of the emotional undertone possibly associated with the retrieval of autobiographical contents. Experimental situations of success or failure such as those implemented by Geen (1979), Seta and Hassan (1980), Sanna and Shotland (1990), or Mash and Hedley (1975) are so many self-related events retrievable in the presence of others. As a function of their emotional undertones, positive or negative, these events are liable, at the time of retrieval, to affect the participant's attention ability with regard to the task to be accomplished (cf. e.g. Ellis & Ashbrook, 1988; Ellis, Thomas, & Rodriguez, 1984).

In short, the perspectives sketched out herein with regard to the problematics defined in Chapters 1 and 2 should allow us (1) to determine the efficacy of the presence of others, under its various paradigmatic forms, as an index of retrieval of self-related knowledge; (2) to evaluate the influence of this retrieval, eventually accompanied by its emotional component, on the participant's attentional ability at the time of the carrying out of the task; and (3) to grasp the part allotted to the involvement of the temporal and autobiographical dimension of the social individual in the determination of his/her behaviour in the presence of others—and that outside the behaviourist framework customarily invoked in connection with this matter in the specialized literature.

CONCLUSION

In keeping with the previous chapters, this chapter about social facilitation encourages us to bear in mind the idea of, if not a determination, at least a close link between the individuals' social history and their cognitive constructions and productions in the presence of other people. According to us, however, this presence, which is by definition a social element, is not enough to endow the individual with a social dimension. It is necessary to take into account the very historicity of his/her cognitive system.

This is what is also suggested in certain researches designed to study "social loafing", a phenomenon often examined in relation to the social facilitation literature (Harkins, 1987; Huguet, 1992; Sanna, 1992). Dedicated to the study of the detrimental effects of working in groups both on effort and on performance, the studies on social loafing will be the theme of the next chapter.

CHAPTER SEVEN

Social-cognitive regulation in co-working group contexts

Human activities often correspond to collective actions in which the contributions of each member are combined into a single group product. Hence the amount of effort exerted by each participant is difficult to evaluate. This is typically the case in collective problem-solving tasks, as in a wealth of other tasks pertaining to various types of organizations (administration, education, industry, etc.). The complete list would be impossible to spell out here, but it certainly matches the widespread belief that working in groups is much better than working in isolation. To reason, to plan, to solve problems as a group: not only are these activities perceived as powerful means of socialization, and hence of social integration (in school, in the company, etc.), they are also thought to enhance each participant's effort and productivity to the benefit of all. Unfortunately, things don't always turn out as well as expected.

In a not too recent book, for instance, the economist Olson (1966/1987) supported the idea that as group size increases, individual effort and initiative decrease: "When the number of participants is important, the average person is fully aware that his/her own efforts will not carry much weight in the outcome of the discussion and that, furthermore, he or she will be touched in the same way by the decision taken collectively, whether or not he or she has exerted efforts to solve the problems at stake. Therefore, average participants will not study the issues as closely as they would if they were alone in the decision-making process. Collective decisions thus seem to be public property as far as the participants (and others) are concerned and each person's own contribution will decrease as the group size increases" (pp.75–6, our translation).

Olson's idea receives experimental support nowadays through the social loafing paradigm (see Huguet, 1995; Karau & Williams, 1993; Shepperd, 1993 for reviews): individuals tend to exert less effort collectively than individually (the social loafing effect *per se*) regardless of the nature of the task (physical, perceptual, intellectual). Admittedly, individuals are not really led to cooperate with each other in this particular paradigm, where the collective work condition is typically reduced to a co-action situation stripped of all communicative interactions. Either the number of real or imaginary co-actors varies or the group outcome is defined as the average of each participant's contribution, with apparently no chance of identifying individual performances.

For this reason, social loafing should not be contrasted with the beneficial effects of cooperation and sharing of viewpoints in collective situations (e.g. our section on related group phenomena). The phenomenon referred to here points to a pessimistic view, not of collective work *per se*, but of what such an activity necessarily entails: co-action. We cannot fail to notice its paradoxical aspect compared to the beneficial effects of co-action in the social facilitation paradigm (Chapter 6). This paradox is further strengthened in the light of a recent meta-analysis of 78 studies (Karau & Williams, 1993), which shows that social loafing generally occurs in simple tasks, independently of the motor, perceptual, or cognitive nature of the activities in play.

Although Olson's remarks seem difficult to dispute in the light of the studies evoked here, they nevertheless identify only some of the factors at the root of social loafing. This phenomenon, unexpectedly discovered by Ringelmann as early as 1913 (cf. e.g. Kravitz & Martin, 1986), is of the utmost importance for the present book mainly because it is also particularly sensitive to individuals' personal and cultural experiences. The specialized literature, however, has not really looked into the meaning of such an observation. It is true that significant progress has been made on social loafing in the past fifteen years. However, because of its tendency to neglect the autobiographical and cultural dimensions of the human being, the literature on this phenomenon has furthered the study of explanatory mechanisms apart from their social conditions of development. In other words, these studies were conducted according to an epistemology that reduces individual products and functionings to the expression of fundamental human motivations in isolation, outside the social institutions and dynamics in which the individual is inevitably placed.

Although at the heart of this chapter, the studies on social loafing will not be examined in detail. In keeping with our goal in Chapter 6, we will underline certain particular facts that will help us to understand further how relatively simple social arrangements contribute towards the regulation of individual effort and performance. After having described the studies on social loafing, we will put forward a view according to which this phenomenon in fact pertains to the social regulation of cognitive performances.

THEORY AND RESEARCH ON SOCIAL LOAFING: A BRIEF HISTORY

The various studies on social loafing stem from Ringelmann's (1913) researches. Without thoroughly describing either the former or the latter, the first part of this chapter will nevertheless allow the reader to acquire at least a superficial acquaintance with them.

Ringelmann's Original Results

As a member of the French Agricultural Society and Professor of Rural Engineering at the National Institute of Agronomy, Ringelmann (1913) studied the efficiency of different modes of human work (studies conducted between 1883 and 1887). As simple as Triplett's (1898) motor task, one of the exercises given by Ringelmann to his students consisted in pulling as hard as possible on a rope 5 metres long (a tug-of-war), both individually and as a group. Results showed an inhibiting effect of the group on individual performances. Specifically, the total strength of the traction produced collectively was always inferior to the sum of the performances accomplished in isolation. Moreover, this inferiority was inversely proportional to the size of the group. It reached 7% in the case of pairs and 15% in the case of trios, and varied between 23% and 77% in groups composed of four to fourteen people! Wrongly perceived as merely an expression of interindividual motor coordination difficulties (Ringelmann, 1913; Steiner, 1972; Zajonc, 1966), such results were ignored for a long time in the psychosocial literature.

The Modern Period

Ingham, Levinger, Graves and Peckham (1974) rekindled the interest in Ringelmann's results. As early as 1974, these authors showed that the effort measured by the rural engineer is less when the participants believe they are contributing to a collective action, carried out by fictitious groups, as opposed to when they are certain they are acting alone. There followed from this observation a spectacular outburst of studies devoted to what is called "social loafing" (Latané, Williams, & Harkins, 1979). This expression underlines the strictly psychosocial component of this inhibition phenomenon (Huguet, 1995). Social loafing, in fact, designates the tendency for people to exert less effort collectively than individually, setting aside all intra-group coordination difficulties. Moreover, it is a general phenomenon, as one can judge by social loafing's insensitivity to the nature of the activities in play. Controlling for coordination loss, subsequent researchers indeed demonstrated performance decrements on collective physical tasks, such as clapping and cheering (Latané et al., 1979), pumping air (Kerr, 1983; Kerr & Bruun, 1981), and swimming in a relay race (Sorrentino & Shepperd, 1978; Williams, Nida, Baca, & Latané, 1989). Performance decrements in individuals working collectively also occurred with both perceptual and cognitive tasks, such as visual vigilance (Harkins & Petty, 1982; Harkins & Szymanski, 1988, 1989; Sanna, 1992), solving mazes (Jackson & Williams, 1985), evaluating essays and job descriptions (Petty, Harkins, & Williams, 1980; Weldon & Gargano, 1988), impression formation (Martin, Seta, & Crelia, 1990), reacting to proposals (Brickner, Harkins, & Ostrom, 1986), and generating uses for objects (Harkins, 1987; Harkins & Jackson, 1985; Harkins & Petty, 1982; Harkins & Szymanski, 1989).

Well-accepted, the empirical reality of this particular type of social inhibition is however still confined to simple, boring tasks, or to tasks devoid of any personal social meaning (Brickner et al., 1986; Sanna, 1992; Williams & Karau, 1991), to such an extent that the inhibition in question, which is not transcultural, was perceived as such in the studies addressing this kind of task (see Gabrenya, Latané, & Wang, 1983). But we will return later on to this important issue, which reveals difficulties of a more theoretical nature.

One must indeed recognize that not a single explanation currently allows one to integrate the whole set of results in the field of social loafing. In this field, as in the area of social facilitation, most hypotheses explain only one aspect among others of the phenomenon observed, and not the phenomenon in its entire complexity.

Explaining Social Loafing

When the contributions of each member of a group are combined, the group output does not point to any given individual effort. Impossible to identify, this effort cannot be reinforced in any way, neither positively nor negatively. This is why, according to Latané et al.(1979), individuals tend to neglect collective work by comparison with more individual tasks.

Social loafing indeed disappears when performance is individually identifiable (Williams, Harkins, & Latané, 1981). More than social visibility as such, however, it is the very exposure to an external evaluation (Harkins, 1987; Harkins & Jackson, 1985), perhaps even to a self-evaluation (Harkins & Szymanski, 1988, 1989; Szymanski & Harkins, 1987) that seems decisive. Whether they entail a personal or a categorical comparison, the anticipation of self-related evaluative events eliminates social loafing. We should not, however, rest on these findings.

As noted by Comer (1995), "it is not simply the potential for evaluation, but actual evaluation itself that is operative in real groups" (p.658). It would thus be unwise to keep on studying social loafing outside the individual's working group experiences. Indeed, the meaning given to the presence of others and to its evaluative potential (cf. Chapters 3 to 6) can depend on the history of various intra-group relations, evaluations, and comparisons.

It is true that social loafing is not exclusively a matter pertaining to this potential. In keeping with Olson's (1987 [1966]) intuitions, this particular type of social inhibition also depends on the perceived probability that somebody else will efficiently handle the problem or the task in hand (Weldon & Mustari, 1988). For this reason, the inhibition in question may very well disappear in cases where individuals perceive their own efforts as unique or as non-redundant when combined with those of others (Harkins & Petty, 1982; Williams & Karau, 1991; Williams, Karau, & Bourgeois, 1993). But here also the individual's experience may play a part.

As suggested by Comer (1995), the perceived uniqueness or redundancy of the personal contribution to the collective task largely stems from the history of intra-group relationships. In the end, it is this history that allows individuals to perceive themselves as more or less competent than their team-mates and thus to compare themselves more or less favourably with them. It is also in the light of these interpersonal comparisons that individuals interpret the division of labour in their team or group. Social loafing can result from the fact that this division is deemed unfair by the participants (Jackson & Harkins, 1985; Kerr, 1983; Kerr & Bruun, 1983).

In the literature on social loafing, however, the meaning given to collective activity is never seen as the product of a personal social history and, hence, as the consequence of evaluation and comparison episodes encountered by the individuals in their group(s). Nevertheless, it is quite possible to imagine that some and even all of the parameters referred to earlier (viz. the perceived evaluation of one's own contribution, the perceived redundancy of one's contribution, or the dispensability of one's effort) are dependent upon this history.

Of course, one might cringe in the face of the very real difficulties connected to the empirical integration of the temporal and auto-biographical dimensions of the social individual (Comer, 1995). But this is no doubt not enough to explain the specialized literature's indifference towards this dimension. In our opinion, this indifference gives expression to the implementation of a dominant epistemological approach. In this approach, the antecedents of behaviour seem to fall into the realm of fundamental human motivations (self-evaluation, for example) outside the history of the social dynamics in which the individual is *necessarily* involved. The question is not one of knowing whether these dynamics can be integrated in the laboratory or not, but rather to what extent they participate in the very unfolding of the motivations and behaviours in question. The stake is high. The fact of neglecting the history of intra-group relationships (perhaps inter-group relationships as well) inevitably leads to an understanding of cognitive functionings constructed outside their social conditions of development. Hence we find, in the social loafing paradigm as well as in other other paradigms of experimental social psychology, an inevitable and exclusive tendency to invoke explanatory mechanisms assumed to be universal and autonomous with regard to the individual's social experience (e.g. Huguet, 1995; Monteil, 1995).

A case in point is the social impact model (Latané, 1981) often deemed one of the most integrative in this field. According to this model, (1) the influence, or total impact ("î") of the real or imaginary presence of others on behaviour is a multiplying function of the social status or of the competence ("S"), of the spatial, perhaps even temporal ("I"), proximity, and of the number of source people ("N") facing the target individual at a given moment, that is to say $î = f(SIN)$; (2) any other source person to whom the individual is confronted increases the total impact (î) according to a power function of the type $î = f(sNt)$, where s is a constant and t an exponent less than 1 (generally close to 0.5); and (3) when the individuals form a common target of influence, in other words when only the group product is identifiable, $î = f(1/SIN)$. In other words, Latané propounds the view that the group can only weaken the experimental impact (by division of this impact), and increasingly so as the size of the group increases. This would not be the case for co-action in the social facilitation paradigm, where individual performances remain identifiable. The experimental impact can thus be separately expressed for each target of influence.

Criticized for maintaining a complete silence on the strictly processual aspect of the phenomena observed (for example by Mullen, 1985), social impact theory does have a remarkable predictive value with respect to the size of the group effects, such as social loafing for instance (see Latané, 1981; for a recent and dynamic version of this theory see Latané, 1996; Latané, Nowak, & Liu, 1994). This very perspective, however, by reducing the individual to the state of a mere "answering machine", does not satisfactorily account for the influence of self-evaluation and of social comparison. More generally, this theory clashes with the results testifying, in the social loafing paradigm, to the weight of the individual's meanings and representations with regard to the task, to the co-actors, and perhaps even to the entire collective situation.

SOME NEGLECTED POINTS

Often neglected as such in the specialized literature, certain results encourage one to integrate the individual's social and cultural history in order to account for the phenomenon at the heart of this chapter.

The Role of One's Social Experience

As suggested earlier, very few studies, among those devoted to social loafing, integrate the role of one's own experience of evaluation and social comparison. Specifically ascribed to interactions or events encountered by the individual in a working group, this experience is in fact totally ignored. In the best of cases, the expectation of a self-related evaluative event is induced in the laboratory in the context of circumstantial groups. The participants, however, are never actually confronted with this event, making it impossible to assess the possible behavioural consequences. Except if one happens to consider that individuals systematically ignore daily evaluation and comparison episodes, there definitely *is* something awkward about this, the more so in that the influence of such episodes is clearly expressed in the social loafing paradigm, provided, of course, that it is not deleted right from the beginning. Here is a striking example of such an instance.

In the preliminary phase of a study conducted by Sanna (1992, Study 1), participants carried out a visual vigilance task and were to evaluate themselves by means of an upward versus a downward comparison based on the number of target-signals correctly detected by peers. Then a new vigilance task was assigned. This next task was accomplished in a co-action versus a collective work condition where individual outputs were pooled to form a single group product. Of course, only the first of these two conditions triggered the expectation of an evaluation of the individual score by the experimenter. Let's briefly check the results.

A differentiated effect of the working context (co-active versus collective) was observed as a function of the type of comparison presented to the participants. Those faced with a downward comparison (expectation of success) were less effective in the collective situation than in the co-action situation. The participants faced with an upward comparison (expectation of failure), on the other hand, generated more errors (misses and false alarms) in the co-action situation than in the collective situation. In other words, the behaviours observed in this study did not depend on the situation and on its evaluative potential *per se*; rather it was dependent on the particular relationship between the individuals and this situation as a function of their comparison experience with regard to the task in hand. This is pretty different and points, once again, to the extreme dependency of current behaviour on the participant's history, the more so when one considers the minimal nature of the comparisons manipulated by Sanna. As opposed to those operationalized in the studies previously presented (for example, those in Chapters 3 and 4), Sanna's comparisons were indeed non-recurrent, and referred to a task that was no doubt new to the participants.

It is true that the weight of the participant's past experiences is completely disregarded in the author's explanatory standpoint. In this standpoint, performance essentially derives from self-efficacy estimations (understood through the meaning given them by Bandura, 1986) coupled with the evaluative potential of the work context. But don't these estimations and self-beliefs stem from the very experiences of comparison encountered by the participant in the study in question? Does this not appear as an invitation to operationalize more systematically the dimensions linked to the personal social history in the social loafing paradigm?

By testifying to the moderating role of "constructed" (Goethals, Messick, & Allison, 1991) or "self-generated" social comparisons (Suls, 1986), our own inquiries into this paradigm (Charbonnier, Huguet, Brauer, & Monteil, 1998; Huguet, Monteil, & Charbonnier, 1995b, 1996a; Huguet, Charbonnier, & Monteil, in press, a) contribute to justifying this interpretation.

The Role of Self-generated Comparison

Sheer mental constructions or reconstructions of events experienced by the individual, self-generated comparisons inevitably partake in a definition of the self and constitute an important aspect of the individual's social experience (cf. Chapter 2). Hypothetical by nature, these comparisons are indeed liable, as is suggested by Suls (1986), to "short-circuit" and overdetermine the self-referred knowledge stemming from actual evaluation and social comparison experiences. This naturally leads to Goethals et al.'s (1991) distinction between "realistic" and "constructive" social comparison: "Realistic social comparison entails self-appraisal based on using and analysing actual information about social reality. Constructive social comparison entails self-appraisal based on 'in the head' social comparison based on guess, conjecture, or rationalization concerning social reality, often believed, and often self-serving" (p.154).

This second type of comparison imparts to individuals a tremendous ability to overestimate their own skills compared with those of others (the "self-uniqueness bias"). As we will see in the studies evoked hereafter, these overestimations of self clearly contribute to the regulation of cognitive functioning in co-working group situations.

Experimental Illustrations

In the preliminary phases of several studies (Huguet et al. in press, a), close to 600 participants answered the social comparison questionnaire devised by Josephs, Markus, and Tafarodi (1992). Participants were to state what appeared to them their best overall skill and to point out their best ability in various fields (sports, academic skills, etc.). Each time, they assessed the personal importance allotted to the skill or capacity in question as well as the percentage of university students sharing the same skill. The lower this percentage, the more the participant described him- or herself as superior to the reference group. The participants characterized by a strong versus a weak feeling of superiority were then faced, in a co-acting or a co-working group situation, with the carrying out of tasks typically used in the social loafing paradigm. The co-actor (co-action situation) or the partner (co-working group context) was always a representative of the reference group (group of students) mentioned in the pre-experimental questionnaire.

For instance, in one of our studies, participants were to generate, in a limited time, as many possible uses for a familiar object for which it was either easy (simple task) or difficult (complex task) to come up with novel ideas. In the co-action situation, the participants expected the experimenter to evaluate their personal performance. This expectation was eliminated in the collective situation, where only the group output (for example, that of a pair) was seemingly evaluated.

As expected, the masking of the individual production inherent in the collective work situation inhibited the efforts of the participants motivated by an interpersonal differentiation process (those high in self-uniqueness). Faced with a simple task, these participants appeared less productive collectively than individually (showing signs of social loafing). Faced with a difficult task, however, these same participants behaved in the opposite way. They were more productive collectively than individually. Probably convinced of their superiority in comparison with "the majority", including their own partner, the participants endowed with a strong feeling of uniqueness could, in the case of a difficult task, view their contribution as necessary to the success of the group taken as a whole. Possibly perceived as yet another proof of their "unique" abilities, this collective success required them to work harder collectively than individually. (This effect, as social loafing, did not occur in participants low in self-uniqueness.)

Could the community or the group therefore benefit from the interference of individualistic behaviours in the quest for uniqueness? We have tried to answer this question through other studies. In one of them, the participants, left in the dark for 40 minutes, carried out a visual vigilance task. Particularly difficult, this task consisted in the detection of 44 $1mm^2$

light signals presented at random on a computer screen amongst 14,000 same-size distractor-signals to be ignored. Endowed, as in the previous studies, with a strong versus a weak feeling of uniqueness in more general dimensions of comparison than the dimension related to the task in hand, the participants worked individually or collectively.

In agreement with the literature on sustained attention and vigilance (Parasuraman & Davies, 1985), the percentage of correct detections decreased over time in two conditions: in the co-action condition for participants high in self-uniqueness, and in the collective working condition for the others. In other words, while the participants low in self-uniqueness showed signs of social loafing, those high in self-uniqueness appeared, once again, more effective collectively than individually (a result rather in contradiction with the ideas circulating in various organizations).

In this particular study, more than the collective situation *per se*, it was in fact the relationship between the participants and the situation, according to self-beliefs derived from subjective social comparison experiences, that came to explain the variations in performances. By encouraging the appearance of a more or less strong feeling of self-uniqueness, the hypothetical comparisons manipulated in this research led the participants, in keeping with those selected in the previous study, to allot a special meaning to the collective situation and to its evaluative potential. For this reason, people's social experience, whether it corresponds to real and/or to more symbolic events, should no longer be neglected in the social loafing paradigm. As a matter of fact, other studies, conducted from an intercultural perspective, point in the same direction.

Social Loafing as a Cultural Phenomenon

Markus and Kitayama (1991) observed that the phenomena of individual performance regulation by the presence of others are not separable from the individuals' cultural affiliations. This observation, which has not been picked up in the field of social facilitation, makes perfect sense in relation to the literature on social loafing. The reality of this phenomenon, indeed, raises no doubts in individualistic cultures such as North America. On the other hand, it appears more questionable in societies that value collective action and group membership, such as China. Gabrenya, Wang, and Latané's (1985) studies are, in this respect, enlightening.

Individualism and Collectivism

In the studies conducted by these authors, North American and Chinese (from Taipei, Taiwan) students were given an auditory vigilance task in the presence of a co-actor. Wearing headphones, the participants were asked to detect target tones working individually or in groups, in which case individual outputs were pooled. Social loafing appeared only in American male students. An opposite phenomenon, called "social striving" (see also Kerr and Bruun, 1981) was observed in Chinese male students. Placed in the collective condition, the former generated a performance equal to 85.2% of the performance produced individually. This percentage reached 110.1% for the latter!

How can one explain these results, which show that social loafing is definitely not a cross-cultural phenomenon, without referring to the social experience of individuals? One is naturally led to integrate this experience in order to explain the phenomena in question. Once again, it is truly the relationship between the participant and a given situation, according to this participant's particular social experience, more than the nature of the situation, that appears decisive. The problem is finding how to tackle this relationship.

According to Gabrenya et al. (1985), the interiorization (understood as a private acceptance) of the dominant cultural norms and values on both sides explains these findings. From this vantage point, "Chinese may have strived when Americans loafed because they felt that the benefit of the group was important and deserving of their efforts, perhaps more so than their own benefit" (p.236). Of course, nothing proves, in this study, that the norms and values really were interiorized. Moreover, the cognitive status of this interiorization should be defined. As is suggested by the authors, the implementation of a self-presentation strategy aiming at conforming with the cultural expectations in use, despite discrepant private beliefs, can also account for the effects observed. The fact none the less remains that, in both cases, the individuals' social experience really *is* involved.

Applied to the intercultural phenomena described by Markus and Kitayama (1991), the effects in question can also be understood as the expression of differing conceptions of individuality. Whereas the self appears as the number one component of individuality in societies such as North America (or Western Europe), on the other hand the self-to-others relationship prevails in collectivist societies such as China, Korea, or Japan. The more specific involvement of these two conceptions of individuality is furthermore noticeable in the fields of cognition, emotion, and motivation (as shown by Markus and Kitayama, 1991). For this reason, it is possible to consider phenomena such as those observed by Gabrenya et al. (1985) as the consequence, not of collective work *per se*, but of the cognitive context resulting from the activation, by the presence of others, of self-related information ensuing from the individual's repeated confrontation with a particular cultural model. When this confrontation is recurrent enough, it may lead to the development of authentic "independence" versus "interdependence" self-schemas, to take up the terminology suggested by Markus and Kitayama (1991).

Admittedly, this terminology might not be the best. The explanation provided by Gabrenya et al., is closer to the distinction put forward by Triandis (1989) between individualistic and collectivistic self-directions: "individualists give priority to personal goals over the goals of the collectives; collectivists either make no distinctions between personal and collective goals, or if they do make such distinctions, they subordinate their personal goals to the collective goals" (p.509). But it is no doubt difficult to settle the question.

Neither mutually exclusive nor totally redundant, both perspectives are today at the heart of an intricate debate (Kashima, Yamagishi, Kim, Choi, Gelfand, and Yuki, 1995). The more so in that other oppositions close to those evoked earlier (the independent vs. the interdependent self; the individualistic vs. the collectivist self) are frequently used to describe the psychological and behavioural specificity of men and women, despite the necessarily intracultural status of gender. To be even more precise, you have no doubt not been oblivious of the fact that gender also seems to play a part in the emergence of social loafing.

The Role of Gender

In agreement with Gabrenya et al.'s (1985) results, Karau and Williams (1993) noticed that social loafing is stronger in studies using an almost exclusively male population in comparison with studies using a mixed population or a population of female participants only. This observation does not come as a total surprise given the literature on feminine psychology (see Chodorow, 1978; Gilligan, 1982; J.B. Miller, 1986). However, the fact that men and women behave differently in the social loafing paradigm is not to be taken lightly.

Indeed, by definition, the category of gender assigns a set of norms and values linked to the way social systems understand and situate men and women in society. Therefore, the behaviours observed in the presence of others are not merely a mirroring of responses linked to the source of variation manipulated in the experimental field. They are also, and mostly, a response that has its origins in the individual's socio-categorical affiliation. This categorical ascendency is rendered possible by the activation or the mobilization of the norms and values internalized by any given individual with the passing of time (Huguet & Monteil, 1994, 1995, 1996).

Fundamental components of social life, the norms and values linked to gender are indeed integrated very early by the individual (Bauer, 1993). Among these norms and values, the self–other differentiation obviously takes on more importance for men than for women (cf. e.g. Lorenzi-Cioldi, 1988; Markus & Kitayama, 1991). Made public, this differentiation, when it is favourable to the self and simultaneously detrimental to others, even seems to be rejected by women (Daubman, Heatherington, & Ahn, 1992). It was thus plausible to expect, despite a more complex link in reality between gender and social differentiation (Huguet, Charbonnier, & Monteil, 1995a), that the possibility of both a public and a positive individuation facilitates the performance of male participants, while inhibiting that of female participants. All this in relation to what the participants are capable of in a situation of anonymity. Let's see.

In Huguet and Monteil's (1995) study, sixty-four 10- to 12-year-olds of both sexes were to learn, while in a holiday camp, a complex figure independent of the gender variable (this figure is the same as the one referred to in Chapters 3 and 6). The learning and recall phases, presented as a play activity, were systematically carried out in the presence of three same-sex co-actors. Almost unknown to the participants at the time of the study (which took place at the beginning of the stay), the co-actors were people with whom participants expected to mingle for at least twenty days. A performance feedback, given following an introductory bogus task, led each participant to expect, unbeknownst to their co-actors, the best possible score on the experimental task. In one half of the groups, participants were informed that individual scores would later be widely publicized in the holiday camp, thus entailing a social comparison. Conversely, for the other half, participants were told that the scores would not be disclosed, except if the participant decided otherwise. Finally, before the carrying out of the task, participants predicted in private their own level of performance and the mean performance of the three co-actors. They also indicated to what extent a public comparison of scores seemed avoidable in the course of the next few days.

The processing of these purely descriptive data showed that the participants actually did perceive themselves as superior to others and either expected (public condition) or did not expect (anonymity condition) that this superiority would be disclosed through interpersonal comparisons. As expected, the processing of the data concerning the free and immediate recall of the complex figure showed that boys performed better in the public condition than in the anonymity condition. The opposite effect was found in girls (see Fig. 7.1). In short, the possibility of a public interpersonal differentiation favourable to the self increased the boys' performance while decreasing the girls'.

No doubt capable of diverse interpretations, these results suggested that, throughout their social history, individuals develop self-knowledge and knowledge about personal situations through the body of norms and values that structures their environment and their group, or more widely their social belonging. Therefore, either explicitly or implicitly activated, this knowledge appears to be involved in the unfolding of certain cognitive

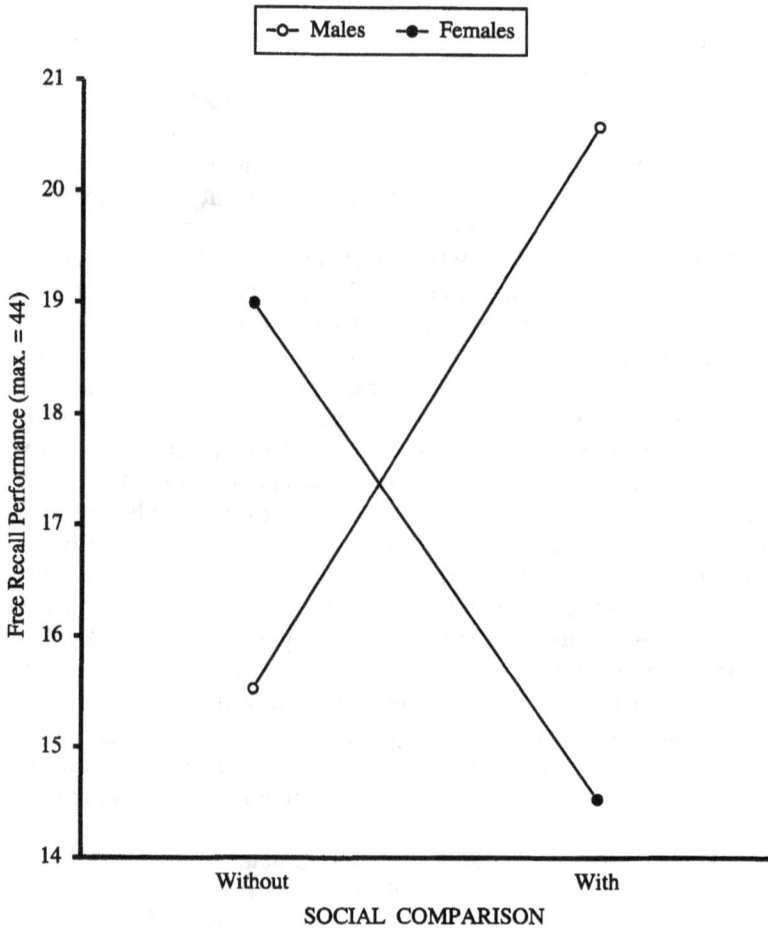

FIG. 7.1   The gender × social comparison interaction on performance.

activities. Although these results were obtained outside the social loafing paradigm, they nevertheless reinforce the results that testify, in this paradigm, to a differentiated sensitivity between men and women towards evaluation and social comparison (see Karau & Williams, 1993). More specifically, our results suggest that women may, in cases where they expect to be superior to their co-actors, exert less effort individually than collectively in order to avoid an interpersonal comparison detrimental to others.

In short, the studies recorded in this chapter offer additional reasons for including individuals' social experience more systematically in explaining their cognitive productions and constructions in the presence of others.

Related Group Phenomena

It is important at this point, in order better to grasp the specificity of the researches at the heart of this chapter, to invoke briefly the link between these studies and those designed to examine the effects of social interaction and cooperation between peers on individual performance and mental development (Doise & Mugny, 1984; see Monteil, 1994 for a review). As the influence of collective work on performance is central to both sets of problems, one may reasonably wonder about their possible connection. However, two major distinctions, one methodological and the other more theoretical, render this connection quite difficult.

Firstly, as we have indicated in this chapter, the study designs called upon on either side are very different. Deliberately shunned in the social loafing paradigm, communicative interaction forms the subject of systematic manipulations in Doise and Mugny's (1984) post-Piaget studies as well as in those linked to cooperative learning (Johnson and Johnson, 1991; Slavin, 1983). Therefore the phenomenon often recorded in these other schemes of research, namely the beneficial influence of co-working groups on performance, is more a reflection of the action of "socio-cognitive confrontations" made possible by the presence of others than of the impact of this presence as such and/or of the evaluative meaning linked to it. Interestingly, one can directly compare the respective effects of the presence of others with the clashing of viewpoints stemming from this very presence (Chernick, 1990; Huguet, Chambres, & Blaye, 1994).

Secondly, from the standpoint of the studies on "socio-cognitive conflict" (Doise & Mugny, 1984), it is precisely this clashing between peers that is invoked to explain the effects of social interaction on performance. The "cognitive decentration" (as defined by Piaget) and the coordination of one's own viewpoints with the contrasting views of others constitute the explanatory mechanisms. In this framework, the interest of the notion of "socio-cognitive conflict" lies in the fact of underlining the social origin of the cognitive decentration conceived as an internal mechanism of regulation of individual cognitive functioning (Monteil and Chambres, 1990).

Social Interaction and Development of Cognitive Skills

As is underlined by Doise (1982), in the thirties certain social psychologists, among the most prominent of the time, arrived at the conclusion that the origin of intelligence was social. Mead (1934), for instance, referred to as a social behaviourist although he was unsatisfied with the Watsonian naturalist and mechanist viewpoint, tried to develop a conceptualization that would understand mental life as the consequence of integrated processes of social life. In that sense, it was more a question of

trying to free psychology from a much too exclusive focus on the individual: "Thought not only involves communication ... but also the production in the individual of the very reaction he provokes in others... One
partakes in the process that the other individual carries out, and one guides
one's actions from this participation" (Mead, 1963, p.63, our translation).

To this conception of a social genesis corresponded, around the same
time (1928), Piaget's ideas on autism as the outer limit of egocentric
thought, on social constraint as the producer of conformism and lack of
autonomy, on cooperation as the producer of reason and on social life as a
necessary condition of the development of logic. Didn't Piaget view discussion as the sinews of verification? According to the Genevan episte-
mologist in those days, social exchange was a major factor in one's
development. Later, Piaget was to abandon this perspective, choosing to
link thought operations to action coordinations rather than to social
transmission. However, these actions are conceived as both individual and
collective.

To the elements briefly recalled here, we must add the studies of
Vygotsky who, in 1930 (Vygotsky, 1978), concluded that the development
of thought proceeds from the social to the individual. The Soviet psychologist suggested a new perspective: the individual's intellectual
development would be the result of the change from an interindividual to
an intra-individual functioning. It is thus obvious that this body of concerns
integrates social interactions with the study of cognitive functioning.

Mead's studies did not lead to experimental researches, and information
about Vygotsky's work was not available; but this is definitely not the case
with Piaget's thought. Indeed, there are numerous studies stemming
directly from his work. However, a closer look shows that they are more
concerned with underscoring the parallelism between cognitive and social
development. The same applies to the studies that start out with a cognitive approach in order to tackle the development of social intelligence.
Thus Damon (1977) examines the unfolding in children (between 4 and 8
years old) of notions such as distributive justice, friendship, etc., and
verifies the existence of a link between the levels reached in this field, for
example the concept of justice, and those reached in Piaget-type cognitive
tasks. The conclusions of these studies are still inscribed in a correlational
perspective between cognitive and social development. Obviously, these
researches do not concern themselves with a possible impact of social
interaction on the development of intelligence.

It is through the Genevan school of thought, described earlier as post-
Piaget, that we may observe a shift in problematics (Doise & Mugny, 1975,
1981). Social interaction is perceived as the privileged site for the
individual's cognitive development. The notion thus put forward suggests
that intra-individual cognitive coordinations are set into place from the

interindividual coordinations. Two propositions follow (see Doise, 1991 for a summary).

Firstly, it is through the coordination of their own actions with those of others that children are led to construct cognitive coordinations that they are not as yet able to make individually. The experimental work here consists in showing that participants, at a given developmental level, succeed in carrying out motor coordination or spatial transformation tasks if they are given the opportunity to accomplish them with someone else (peer or adult), even though they cannot perform these tasks on their own. Secondly, participants who have accomplished these tasks with someone else are then able to carry them out on their own.

These propositions have been experimentally founded by the cooperative play paradigm (Doise & Mugny, 1975). A task requiring independent motor coordination, and feasible only if the actions accomplished result from an adequate correspondence between the partners' cognitive operations, is performed by 7- to 9-year-old participants. The children must use a pulley to move a mobile, attached to it with strings, in a certain predetermined direction. Two children manipulate one pulley each, whereas in the control condition, only one child uses the pulley. Results show the superiority of participants working in pairs, but only from the age of 7 and up. The social interaction situation truly *does* produce a better coordination of actions at an age where participants are still developing their systems of coordination, but is less beneficial for participants who have already mastered the necessary coordinations. The anteriority of interindividual coordinations over individual ones is thus demonstrated.

This type of research aims at identifying and defining the social conditions that generate the production of these coordinations. The Genevan researchers put forward the hypothesis that the intensification of a symmetrical interaction between two partners led to use opposing approaches constitutes one of these privileged conditions. On the other hand, the interaction with a more advanced partner is not always a source of progress. By studying how children working in pairs resolve the problem of the Hanoi tower (Richard, 1982), Glachan and Light (1982) observed that children working together do not perform better than children working alone. However, when the child is forced to take into account his/her partner's suggestions, the performances produced collectively become superior.

The results of these various studies lead one to wonder how it is possible to explain the mechanisms by which a social interaction can prompt a surpassing of one's personal cognitive capacities. The answer calls in aid the idea of a centration conflict as a phase of development. The hypothesis that should be submitted to a proof by facts is that of a social opposition of centrations as source of progress. This particular interaction, known by the generic term "socio-cognitive conflict", primarily refers to an interactive dynamic assuming that the participants actively engage in a cognitive confrontation productive of various oppositions and differences of viewpoints. The many experiments conducted (Doise & Mugny, 1981) underscore the fact that the social interaction is more particularly a source of progress "if there is induction of a confrontation between the actors' diverging solutions" (p.81).

As one may notice, it is not a question here of viewing the presence of others according to the meaning given to this presence in this and other chapters. It is obvious, however, that, once again, the interindividual social dynamics exert an influence on individual cognitive activities.

Cooperation and Competition

The study of the effects of cooperation and competition is a traditional theme of research that stems from Morton Deutsch's studies conducted at the end of the forties. There is nothing to add to the definitions put forward by Deutsch (Deutsch, 1949). A cooperative structure implies that one can only achieve one's goal if the other achieves it too. Conversely, a competitive structure means that one can achieve one's goal only if the other fails.

If we confine ourselves to the more salient results found in the relevant literature, which is rich in this particular field, it seems a recognized fact that cooperation produces better results on task performances than does competition (see for example Ames, 1984; Johnson & Johnson, 1989; Slavin, 1985, 1990). For this reason, the implementation of cooperative learning programmes represents one of the possible psychosocial applications in school. This type of learning entails teaching methods in which students work in small groups towards a common goal. As opposed to competitive classroom structures, the main characteristic of cooperative learning is that the success of one student contributes to the success of all. But the fact of merely stating the superiority of the cooperative structure over the competitive structure does not, however, explain how this comes about. The more so in that this superiority seems to apply to a variety of skills: reasoning, memorization, metacognitive strategies, etc.

The cooperative structures must however present certain characteristics to explain their efficacy in cognitive activities and productions. Mainly, they must offer the essential elements of a positive interdependence that includes the perception of a personal responsibility in the participants. Several experimental studies have thus shown that when the part allotted to each person is strictly defined beforehand and different for each, the cooperation loses some of its efficacy. Conversely, a situation in which roles are loosely defined, allowing space for differentiation and interindividual confrontation, provides cooperation with the means of showing its full worth. Indeed, in the absence of pre-defined roles, participants must strive to construct a minimal division of labour. The functional obligation is liable to generate conflicting viewpoints. In line with the studies on socio-cognitive conflict, we know that opposing centrations of viewpoints facilitate cognitive constructions. The cooperative learning structure, by imposing the point of view of others, appears as the structure most likely to facilitate the emergence of opposing centrations, of which the effects are well known. Moreover, this structure maintains an interdependence of functionings and of goals in the partners. Furthermore, several experimental results (e.g. Johnson & Johnson, 1989) point to the fact that in this type of configuration students cognitively organize the material to learn in a more elaborate fashion than when they have to do it for their own sake only. In other words, the necessity of exchanging the cognitive material with others constitutes a basic factor in the development of metacognitive competences transferable to individual learning.

Even on the basis of the avowed superiority of cooperative structures over competitive structures, it would obviously be dangerous to use this as an argument to subject the situations underlain by competitive structures to public obloquy. The more so in that academic practices generally take place in collective settings where social interactions and the distribution of reinforcement and evaluative feedback generate complex social comparison processes, the complexity and modalization of the effects of which we have foreseen. Social comparison is obviously not separate from competitive dynamics. These dynamics, however, must not be confused with the ideological presuppositions that too often saturate their defence to the benefit of competitive education methods of a more proselytizing type. As one of us (Monteil, 1989) has stated elsewhere, we will reassert here that competition must be the sphere of application of acquired competences, rather than the social basis of their acquisition.

From the standpoint of the various elements presented in the course of this and other chapters, it would no doubt be justifiable to wonder by what means the phenomena described or invoked here are in a position to provide possible pointers to action, in particular in the field of education. We will introduce these pointers in the next chapter.

CHAPTER EIGHT

Pointers for educational action

We could not end this book without underscoring the importance for educators of certain experimental results described in the previous chapters. The results recorded in Chapters 3 and 4, in particular, shed a different light on well-established classroom practices, for example the habit of publicly quizzing poor achievers in order to help them improve their performance or their abilities (at least in France). Our results show that, even in the case of positive feedback, social visibility and exposed social comparison may inhibit the performances of this category of students. The widespread idea according to which interpersonal competition, closely linked to visibility and social comparison, necessarily leads to better performances thus appears questionable, to say the least. Competition produces a double effect even in high achievers: either positive or negative, as a function of the nature of the past events activated by the current learning conditions or by the performance situation. Of course, the variables operationalized in our studies (social comparison, public individuation, the prestige of the academic disciplines) only illustrate a small part of the dimensions simultaneously in play in a classroom (Marshall & Weinstein, 1984). The lessons drawn from these studies, however, prompt us to consider the meaning of these dimensions as closely linked to the individual's academic history.

While contributing to a new way of approaching cognition, these lessons are also liable to call into question educators' practical experience of teaching and, hence, the psychosocial literature most likely to enlighten this experience. This is why two additional questions will hold our attention in this final chapter. Firstly, we will see in what respect our results inform the literature on social comparison in school; secondly, we will put forward certain thoughts suggested by our results with regard to educational practices.

SOCIAL COMPARISON IN THE CLASSROOM

According to Levine (1983), three factors are liable to transform the classroom into an extremely favourable environment for social comparison: (1) students' uncertainty about the ever-changing success criteria linked to academic learning; (2) the fact that the performances and efforts accomplished by each student are generally followed by positive or negative reinforcements, and are thus subject to a continuous evaluation; and (3) the high density and accessibility of individuals relatively similar to one's self. Combined together, these factors almost inevitably lead to social comparison (cf. Chapter 2), to the extent that the activity of comparing one's own performances with those of others sometimes emerges despite educational strategies aimed at eliminating it (Crockenberg & Bryant, 1978).

From this stems the interest of a systematic study of the effects of social comparison on attention and on cognitive performances at school (cf. Chapters 3 and 4). This study has, by the way, been widely neglected in the specialist literature. In one of the few studies of these effects conducted with kindergarten children, not only did Santrock and Ross (1975) point to the importance of social comparison as a factor liable to influence attention and performance (in the same way as self-confidence), they also suggested that the subjects' personal social histories (that is, the comparison episodes they had previously experienced) played a mediating role in this influence (see p.197). Unfortunately, this interesting conclusion was not to be followed by other developments. With a few exceptions (Halisch & Heckhausen, 1977; Hokoda, Fincham, & Diener, 1989; Spear & Armstrong, 1978), the aim of the studies on social comparison in school was to identify the causes and the various stages of its development rather than to analyse its consequences for performance. To what conclusions do these, albeit important, studies lead?

Already noticeable in very young children of kindergarten age (France-Kaatrude & Smith, 1985; Masters, 1971; Mosatche & Bragonnier, 1981), the use of social comparison seems to fulfil different goals depending on the level of the individual's cognitive development. In younger children (before the age of 7), social comparison serves in affiliation to and acquisition of the norms and rules in force in the school setting. In older children, it appears to answer a need for self-evaluation (Frey & Ruble,

1985; Ruble & Frey, 1987; Smith, Davidson, & France, 1987). Indeed, younger children do not modify their own self-evaluation even after being made aware of the performances of others (Ruble, Boggiano, Feldman, & Loeble, 1980). This phenomenon seems to be linked to the absence in these youngsters of the inferential abilities required for an optimal use of social comparison information (Ruble, 1983). Consequently, it seems that in very young children self-evaluation depends more on a "temporal" comparison (cf. e.g., Chapter 2) rather than on a "social" comparison. In particular, primary school pupils appear to evaluate their own performances by evaluating them with their past performances rather than with the performances of others (Ruble & Frey, 1987). This literature also suggests that this temporal comparison should be encouraged in older children because of the seemingly detrimental effect of social comparison on self-evaluation with the passing of time.

As suggested by Wood (1996, p.523), however, the fact that the awareness of the performances of others does not always disturb self-evaluation does not necessarily entail an inability to process the characteristics of others with regard to various self-representations—processing that is at the very root of social comparison. It thus seems fair to question the influence of this processing on students' cognitive elaborations and productions, even in the youngest pupils. From our point of view, the influence of the comparisons in which the individual is currently engaged, either freely or under the pressure of the environment, depends for the most part on his/her past experiences. Nothing will prevent (1) the current social comparison experiences, including those occurring at the early stages of cognitive development, from being stored in memory and later retrieved under the right activation conditions; and (2) this encoding and retrieval occurring independently from the individual's cognitive/inferential abilities, all the more so when the nature of the memory operations in play is implicit (cf. e.g. Chapter 2).

This is why it is quite reasonable to imagine that starting early on in the individual's cognitive development, academic performances may depend on an interaction between past and present social comparison situations, thus carrying the sign of the students' academic history, *even in an embryonic state*. For this reason also, the "social" and "temporal" aspects of a comparison cannot so easily be dissociated. Of course, for clarity of thought, one can indeed oppose these two aspects. It seems to us, on the other hand, much more tricky to reduce temporal comparison to a "comparison by default", that is an activity that would set itself into motion only when social comparison is not conceivable because of insufficient cognitive abilities (Suls & Mullen, 1982). It is not a question here of denying the importance of such abilities in social comparison. Rather, it is a matter of investigating the possibility of a simultaneous influence of very young children's past and present social comparison experiences on their academic performances. This as yet has not been tackled by the specialist literature. The results recorded in Chapters 3 and 4 clearly testify to the existence of this double influence in secondary school students. Hence, our suggestion is that one should *remain careful when it comes to managing social comparison in the classroom, whatever the age and the level of cognitive development of the individuals involved.*

A FEW SUGGESTIONS

We have only to remind ourselves of the two questions most often asked in the classroom by a teacher when the time allotted for doing an exercise is up: "Who hasn't finished yet?" and "Who was unable to do the exercise?" Through these seemingly harmless questions, the students are in fact confronted with a "forced social comparison" that, according to the dimensions at work, can be systematically favourable or unfavourable to the student, even without the teacher's knowledge. Repeated tens, perhaps even hundreds of times, during the course of a single academic year, and probably thousands of times during the elementary programme, these same questions will have triggered so many comparison episodes liable to be encoded in memory. "Clandestine" with regard to the teacher's true intentions, these episodes should nevertheless be the target of constant attention, in the same way as attention is naturally granted to the material conditions of teaching and to the efficiency of such-and-such a teaching aid. These conditions and aids are obviously important in the context of the educational goals to be achieved. However, the achievement of these aims may depend at least as much, if not even more, on the comparison dynamics generated by the educational practices and their effects on cognition. It would thus be important systematically to control and plan the social comparison situations almost always associated with these practices. The fact of neglecting these situations only leads to the haphazard creation of socio-psychological configurations with possibly disastrous consequences. Unintentionally to confront students already facing difficulties with a systematically negative comparison while increasing competition inside the classroom can produce at least two harmful effects. Combined together, a negative comparison and competition can worsen cognitive performances (cf. Chapters 3 and 4) and also, as shown in a past study (Santrock, Smith, & Bourbeau, 1976), considerably increase the frequency of aggressive and regressive behaviours (disengagement from the activity in progress, self-centration, and a tendency to isolate oneself). Regularly made to feel inferior to his/her peers, the individual may also engage in a pursuit of "non-comparability" (Lemaine, 1966, 1974), expressed by the production of extraordinary behaviours inappropriate to the academic system, perhaps openly and durably identifying with the most devious individuals in his/her social environment (Kagan, 1990).

As we have seen, however, social comparison can be seen as highly positive in some cases (a phenomenon neglected by the specialist literature), which is another reason to integrate it into basic educational practice. On that subject, the propositions put forward in Chapter 5 with regard to social comparison feedback are of particular interest here.

Social Comparison Feedback

The information given to the students about their performances often takes, explicitly or not, the shape of a social comparison feedback (previously defined as "the involvement of an outside agent who, by providing an evaluation on the performance, the competence, or the status of an individual, places him or her in a situation of comparison to others"). This feedback must also be manipulated with precaution. The propositions put forward in Chapter 5 suggest various steps to take in this manipulation. Basically, this approach consists in creating, through a play on the saliency of the social comparison, the conditions of an optimal processing of the target-information, taking into account the dimensions of comparison involved and the students' past experience in the matter.

The fact of favouring anonymity rather than public individuation, for example, can be particularly beneficial to the learning process in students usually made to feel inferior in such-and-such a dimension of comparison. Paradoxically, the creation of such a context even seems to constitute, in these students, a necessary condition for the expression of a positive comparison feedback effect unusual for them (that is, a success feedback). But in this respect, the teacher usually acts the other way around, making visible the poor achiever's unexpected success (at least this is what happens in France). As we have seen, social visibility is very efficient for students whose success in a given area is the rule rather than the exception. On the other hand, it is inadequate in the case of the poor achiever, in whom success coupled with high visibility can consume a good part of his/her attentional resources to the detriment of the task itself.

It would thus seem naive to think that in a learning situation the mere fact of accumulating encouragements, particularly in the shape of success feedback (legitimate or not) will necessarily lead to the enhancement of the poor achiever's performances. It is still important to take into account the saliency of the feedback and to be careful not to introduce too many unfamiliar events at the same time. In the opposite case, the encouragements may never be followed by any effects save an effect of discouragement of the teacher towards the low achievers. Followed by a lack of attention towards these poor achievers, this discouragement is likely to reinforce academic failure and to foster well-known dynamics such as the "self-fulfilling prophecy" (Rosenthal & Jacobson, 1968; cf. also W.P. Robinson, 1984) or the teacher's ability to change the student's status in the direction of his or her own expectations (positive or negative).

On the other hand, once the poor achiever has thoroughly integrated the success feedback, it can no doubt be coupled with more social visibility. The problem is to know under what shape and, most importantly, at what moment. If indeed public individuation inhibits performance when it occurs too early on (when success is still an unfamiliar situation), one must expect anonymity to produce an identical effect when it lingers on pointlessly (when success represents a now familiar situation).

No doubt we may find in these issues a renewed interest in collective work and cooperative learning. Indeed, by concealing individual performance, collective work, which we dealt with in the previous chapter, may constitute an attractive situation for students experiencing regular problems in such-and-such a dimension of comparison. In line with this idea, Willerman, Lewit, and Tellegen (1960) showed that a dread of failure does indeed encourage pupils to seek collective rather than individual work. But as we have seen, collective work does not only present good points. It can both encourage the good student to loaf, because of its anonymous nature, and incite the poor achiever to take advantage of the competences and efforts exerted by the good student, which in turn represents a good reason for the latter to loaf.

However, as indicated in our own studies (Chapter 7), the anonymity often associated with collective work is not necessarily detrimental to individual effort and performance, in particular in the case of collective tasks deemed difficult to carry out. In that case, it is sufficient that the individual (rightly or wrongly) perceives him/herself as superior to the other members of the group, and thus instrumental towards the achievement of the goal, for him/her to exert more effort collectively than individually. It is precisely when the task to carry out is difficult that collective work becomes relevant, particularly for the poor achiever. From this stems the idea that high and low achievers alike can benefit from collective work under certain conditions.

Indeed, if working with peers (as understood in the studies on cooperative learning) is generally accompanied by a beneficial effect on performance, it also presents the additional interest of collectively confronting the poor achiever with success. And the possibility is not ruled out that this "silent confrontation" with success, when it happens often enough, may lead the poor achiever to ask for more competition and social visibility, just like the good student. Even in a context of success, however, social visibility does not always have the effect expected, in particular when personal achievement implies a negative comparison for others. Certain studies presented in Chapter 7 (Huguet & Monteil, 1995, 1996) indeed show that in such a situation, visibility can inhibit as well as facilitate performance according to the norms and values associated with gender. In the study in question, the prospect of a public individuation favourable to the self enhanced the boys' performance and impaired the girls'. Compatible with the idea, well-established in Western society, that feminine norms and values are oriented towards cooperation and empathy rather than towards competition and domination, these results are not pragmatically neutral. They suggest that the norms and values inherent in the individual's sexual identity should be taken into account during the manipulation of comparison feedback. The internalization of these norms and values in fact occurs early in cognitive development, if one can judge from certain results testifying to the presence of authentic gender schemas even before language acquisition (Bauer, 1993).

In short, the comparison feedback and the degree of social visibility made available to students should become the target of truly strategic manipulations. In this perspective, two other factors are important: the value associated with the dimensions of comparison; and the congruence between the contexts where the individual encodes and retrieves the information supplied by the teacher.

The Dimensions of Social Comparison

Certain studies expounded in Chapter 3 (Monteil, 1988, Study 3) show that cognitive performances vary considerably as a function of the meaning allotted by the students to the dimensions of comparison. Compared to the neutral or less-valued dimensions, the most valued dimensions of comparison force the student to behave as expected by the school system. Thus, the saliency of the comparison depends at least as much on the degree of social visibility offered to the students as on the value associated with the dimensions of comparison. For this reason, the precautions outlined earlier concerning comparison feedback and public individuation of academic performances should be heeded in the case of disciplines involving high social stakes. These disciplines are the most likely to spark the success and failure dynamics described in Chapters 3 and 4.

In addition, one may recall that manipulation of the context of presentation of the task is enough in itself to provoke a noticeable variation in cognitive performances (Monteil & Huguet, 1991). The fact of noticing that the performance of poor achievers is enhanced when the task relates to a less prestigious discipline, such as drawing rather than geometry, suggests than one should work on the "dressing up" of academic disciplines. Frequently undertaken with regard to the characteristics of the task's strictly material environment, this manipulation requires that particular attention should be given to the tasks' evaluative meanings as a function of their corresponding disciplines and academic knowledge.

From this vantage point, it is not necessarily satisfactory to try to increase, through computer science for instance (see Snyder & Palmer, 1986), the play aspect of academic subjects in the hope of making them more attractive and more intellectually productive. A maths problem, whether presented on a computer screen or not, will always be more evaluatively charged than any other school assignment, at least as long as our current cultural models say so (and they are not close to changing!). And as a matter of fact, the use of computers in the perspective presented here has already been widely criticized (Bowers, 1988; D'Attore, 1981; Psotka, 1982).

More generally, it is not certain that the play component of learning systematically enhances performance (Lepper & Hodell, 1989). Many times postulated (Bruner, 1961; Condry, 1977; Deci, 1975; Lepper & Greene, 1978; Singer, 1977), this idea is only confirmed nowadays from a body of correlational studies that are by definition unable to determine the true cause of the effects observed. Do we learn better because the material is deemed attractive, or is the material deemed more attractive because the learning goals are successfully achieved? (Only two recent experimental studies conducted by Parker and Lepper in 1992 provide sound experimental support for the former conclusion.)

It still remains that in our own studies (Monteil & Huguet, 1991), the high achievers performed less well in a more playful context, that of art/ drawing, than when they were faced with a more academic task such as geometry. Nakamura and Finck (1973) also observed that, in good primary school students, the task's "playful dressing up" can promote laxity, thus impairing performance in comparison with what is produced in a more evaluative context. Of course, this type of dressing up *does* enhance the low achievers' performance in our studies. But, as shown in a post-experimental study (Huguet & Monteil, 1992), this beneficial effect was not really due to the playful aspect of the dressing up in question.

In this study, a sample representative of the experimental population selected by Monteil and Huguet (1991) was required to assess (1) the reliability of using performance in specific academic disciplines to evaluate general intelligence; and (2) the cognitive abilities (memory, imagination, thought) required to succeed in these same disciplines. As expected, high and low achievers disagreed over the abilities deemed necessary in the case of less valued disciplines, such as drawing, physical education, sports, or crafts. Only the low achievers, who in general were not running into any difficulties in these particular disciplines (as opposed to the disciplines more valued by the school system), assessed memory, thought, and imagination as abilities essential to these school subjects.

On the other hand, there was very little difference between the two groups with regard to the perceived reliability of subject-based academic knowledge as a measure of intelligence. The less-valued subjects (with the exception of crafts) were deemed the least reliable by most participants. In other words, the low achievers, who are consistently exposed to negative comparisons in the valued disciplines, tried to enhance the dimensions that were favourable (or less unfavourable) to them. However, this enhancement did not go so far as to lead them to question the dominant hierarchy of disciplines with respect to their relation to intelligence. Mathematics and first-language acquisition still remained at the top of the list.

These results are interesting on more than one account. This subtly original conception of the field of social comparison in low achievers contributes to explaining why these students do not systematically base their academic identity on anti-academic values. This phenomenon is in other respects deemed surprising (W.P. Robinson & Tayler, 1989; W.P. Robinson, Tayler, & Piolat, 1990). As was shown by Lemaine (1966, 1974), the redefinition of the social comparison field is an efficient strategy for fighting inferiority situations and safeguarding one's "living space" in a social system that is quite threatening in other respects. But above all, when related to the experimental study described earlier (Monteil & Huguet, 1991), these observations suggest that the beneficial effect on low achievers of the task's playful dressing up depended less on this dressing up *per se* than on the students' past experiences in the reference discipline (drawing) and on the evaluative meanings the participants assigned to this discipline. It is thus important to concentrate on the meanings assigned by the students to the academic tasks and learning rather than only on their play aspect.

Social Comparison as a Retrieval Cue

Finally, in an interactive conception close to Tulving's (1983; Chapter 1) encoding and retrieval processes, it can be expected that the performances linked to the recall and/or to the recognition of a target-information (i.e. the information given by the teacher) depend in part on the compatibility between the contexts in which the individuals encode and retrieve this information. If certain characteristics of the learning contexts, in particular the conditions of comparison, are encoded during the processing of the target-information, then the presence of these characteristics at the moment of retrieval should facilitate performance.

By showing that the students do not manage their current comparison experiences independently of their past experiences on the same matter, the studies described in Chapters 3 and 4 provide evidence to confirm this hypothesis further. In these studies, however, the current comparisons offered to the students seemed to play the role of retrieval cues for past events of reference (that is, social comparison episodes) rather than that of retrieval cues for target-information *per se*. Thus the hypothesis in question still has to be verified. From an applied point of view, the stakes are very high. If indeed it turns out that our hypothesis is correct, the idea that the evaluations and comparisons in play in the classroom are authentic retrieval cues can help to redefine the testing conditions (according to the learning conditions) and eventually to allow a more accurate interpretation of the performances observed in students.

We must recognize that the Tulvinian theory of a "specific encoding" has never taken into account the role of the individual's social environment. Labelled as environmental contexts, only the psycho-physiological and thymic states (i.e. the "internal" contexts), and the individual's physical environment (i.e. the "external" context) have been systematically manipulated. And the effects of these manipulations on memory often proved inconsistent.

For instance, Smith (S.M. Smith, 1979) has tested the importance of carrying out the study of the material to remember and of the recall test in identical physical environments. Participants were to study a list of words in Room "A". Afterwards, they moved on to Room "B", where they were to draw the room as seen from various angles. The participants next moved to a waiting-room, where they remained for a few minutes. Finally, half the participants returned to Room "A" to perform the recall test (same context), whereas the other half stayed in Room "B" for the test (different context). In agreement with the specific encoding theory, the first group (same context) accomplished a recall performance more than 25% higher than that of the second group (different context) (see also S.M. Smith, Glenberg, & Bjork, 1978). In a second experiment, Smith (1979) also showed that, prior to the retrieval phase, the mere evocation of the room reserved for the learning phase was enough to facilitate the recall of target-information. These, and other (S.M. Smith, 1982, 1984) results, deemed capable of being used in academic settings (S.M. Smith & Rothkopf, 1987), have provided support for Tulving's approach.

Other studies (Fernandez & Glenberg, 1985; Saufley, Otaka, & Bavaresco, 1985), however, have shown that, contrary to what Smith predicted, it is in fact impossible to act on memory systematically through the manipulation of the physical environment of individuals alone. Despite a noticeable increase in the statistical power of the test used by Smith (1979), Fernandez and Glenberg (1985) do not detect any advantages in the conditions where the study and the memory tests were conducted in an identical room. Likewise, Saufley and his colleagues (1985) compared several groups of students' mid-term exams results. These students either wrote their exams in their usual classroom or in a different room. No difference in the usual output of these two groups of students was observed. Fernandez and Glenberg (1985) thus concluded that the physical environment is not, as such, necessarily relevant for the individual. We readily concur with this conclusion.

In our perspective, indeed, the contextual components perceived as relevant for the carrying out of the task refer more to the social individual's temporal and autobiographical dimension than to his/her environment as such. We are not denying that an effect of the "physical context" on memory can be obtained in certain circumstances, for instance when the environmental cues are closely linked to the information to be retrieved. However, the connection between the individuals and their current situation, as a function of their previous social experiences, suggest that the context only exists through the involvement of cognitive structures of contextualization, such as those linked to the individual's autobiographical memory. From this vantage point, the "external context" referred to in the specialist literature appears theoretically inadequate. The same applies to the so-called "internal" context. By referring exclusively to the individual's psycho-physiological and thymic states, its use at once disposes of the idea of an autobiographical contextualization process. As we have previously remarked (Chapters 1 and 5), it is far from impossible that certain internal states, in particular emotional states, are in fact indissociable from this very process. This hypothesis is purely and simply ignored by the memory specialists. In short, whether they are labelled as "internal" or "external", the contexts in question are conceptually remote from the "cognitive context of self" invoked all through this book as the consequence of an interaction between the history of the cognitive system, its products, and the current conditions of retrieval.

CONCLUSION

Often incomplete, sometimes highly speculative, these few considerations should nevertheless contribute to a new approach to certain questions linked to academic learning and performance. The fact of acknowledging that social comparison constitutes one of the basic cognitive activities for students implies that all the practical consequences should be drawn. This has not been done in the psychosocial literature. The importance of social comparison has indeed been acknowledged, but the results of the studies conducted in this field have not been discussed from a practical point of view. One can only regret this fact and hope that future studies will develop this applied perspective, without which the results in question cannot take on their full meaning. The aim of this last chapter has obviously not been to fill all the existing gaps. Developed from our own studies, the practical indications provided here almost exclusively refer to the effects of social comparison on cognitive performances. Social comparison also constitutes a powerful determinant of behaviours that in turn appear as extremely important in the academic setting, such as persistence in the activities in progress (W.P. Smith, Davidson, & France, 1987) or affiliation (see Goodenow, 1992; Tesser, Campbell, & Smith, 1984). We can no doubt extract from that literature new arguments in favour of new practices in the field of education.

General conclusion

The experimental underscoring of the social regulation of academic performances is certainly inadequate to extend it *de facto* to cognition in
general. Must we for that reason reject the parallels outlined? In so far as
the book is committed to this idea, the authors cannot honestly decide.

Moreover, since this involves questions linked to the effects of previous
social experiences and knowledge on the perceptions and reactions of
individuals, researchers in social psychology have not waited for this book
to ask them. Numerous studies conducted in the field of social cognition
point to concerns close to those expressed in these pages. The priming
effects on social categorization, the effects of stereotypes on the perception
of people, the effects of the accessibility of a schema in memory for
information consistent or not with this schema or the effects of practice on
efficacy and on the content of social judgement are other examples of this.
Theoreticians and researchers in social cognition do their best, in the
different fields of investigation, to identify the mechanisms that underlie
these effects (see for a discussion E.R. Smith, 1990). They are aimed at
understanding how knowledge previously acquired is liable to become
involved in the processing of a piece of social information. The methodologies and the basic concepts invoked obviously come under cognitive
psychology. Nothing more legitimate. We can't imagine how one can
approach the processing of a social object without calling upon information
processing theory and the studies on memory. It is therefore logical to
envisage, beyond the experimental method, the use of simulation tech-

niques (E.R. Smith, 1990). This evolution tends to erase the boundary between cognitive psychology and social psychology. However, by having the noun "cognition" following the adjective "social", the social cognition trend refers to a specificity based on the likelihood of more complex mechanisms for the processing of so-called social objects than for that of non-social objects.

In this book, our ambition has been to offer a different perspective by shifting the objective, in the optical meaning of the term, from the characteristics of the object to those of the social individual. Of course, in the study of the cognitive mechanisms used in the processing of social objects (others, social groups, the self, etc.), the individual is obviously at the heart of the procedure—otherwise how could one speak of cognition? However, this individual is often understood under a conception that, to use the dominant terminology, reduces him/her to the status of a mere "perceiver". This is why we have chosen to consider the individual in relation to his/her social insertions, in so far as the nature of a socially inserted being is also a universal property of individuals. *In our opinion, it is more the social nature of the individual rather than that of the object that truly defines the social nature of cognition.*

Indeed, if social situations really constitute sources of modifications of cognitive performances and activities, the studies presented in this book seem to indicate that their influence is linked to the individual's very history. Acquired, formed, and used in relationships often implying the self or a representation of self, the knowledge of situations, people, groups, or objects no doubt leads, in the understanding of such-and-such behaviour, to the taking into account of the autobiographical contents it carries. Likewise, in so far as the individual constructs his/her own social component in relation to the presence of others, we cannot consider this component as the mere natural expression of a social state. On the contrary, it is advisable, through social comparison for instance, to understand all its dynamic meaning for cognitive regulation.

Thus, to adopt the point of view of a social psychology of cognition entails the recognition of others as elements of individuals' personal history and, hence, as one of the determinants of cognitive expressions and functionings.

References

Ackerman, P.L. (1987). Individual differences in skill learning: An integration of psychometric and information processing perspectives. *Psychological Bulletin, 102,* 3–27.

Albert, S. (1977). Temporal comparison theory. *Psychological Review, 84,* 485–503.

Allport, F. (1920). The influence of the group upon association and thought. *Journal of Experimental Psychology, 3,* 159–182.

Ames, C. (1981). Competitive versus cooperative reward structures: The influence of individual and group performance factors on achievement attributions and affect. *American Educational Research Journal, 18,* 273–287.

Ames, C. (1984). Competitive, cooperative, and individualistic goal structure: A cognitive motivational analysis. In R. Ames & C. Ames (Eds.), *Research on motivation in education, 1* (pp. 177-208). New York: Academic Press.

Ammons, R.B. (1956). Effects of knowledge of performance: A survey and tentative theoretical formulation. *Journal of General Psychology, 57,* 279–299.

Andrew, R.J. (1974). Arousal and the causation of behaviour. *Behaviour, 51,* 135–165.

Annett, J. (1969). *Feedback and human behaviour.* Harmondsworth, UK: Penguin Books.

Arp, G.F. (1920). Work with knowledge of results versus work without knowledge of results. *Psychological Monographs, 28,* 1–41.

Asch, S. (1946). Forming impressions of personality. *Journal of Abnormal and Social Psychology, 41,* 258–290.

Avrahami, J., & Kareev, Y. (1994). The emergence of events. *Cognition, 53,* 239–261.

Babab, E. (1990). Measuring and changing teachers' differential behavior as perceived by student and teachers. *Journal of Educational Psychology, 82,* 683–690.

Baddeley, A. (1992). *Human memory: Theory and practice.* London: Lawrence Erlbaum Associates.

Balcazar, F., Hopkins, B.L., & Suarez, Y. (1985). A critical, objective review of performance feedback. *Journal of Organizational Behavior Management, 7,* 65–89.

Banaji, M.R., & Crowder, R.G. (1989). The bankruptcy of everyday memory. *American Psychologist, 44,* 1185–1193.

145

Bandura, A. (1986). *Social foundations of thought and action: A social cognitive theory.* Englewood Cliffs, NJ: Prentice-Hall.

Barclay, C.R., & Hodges, R.M. (1990). La composition de soi dans les souvenirs auto-biographiques. *Psychologie Française, 35,* 59–65.

Barclay, C.R., & Subramanian, G. (1987). Autobiographical memories and self-schemata. *Applied Cognitive Psychology, 1,* 169–182.

Bargh, J.A. (1982). Attention and automaticity in the processing of self-relevant information. *Journal of Personality and Social Psychology, 43,* 425–436.

Bargh, J.A. (1989). Conditional automaticity: Varieties of automatic influence in social perception and cognition. In J.S. Ullman and J.A. Bargh (Eds.), *Unintended thought: Limits of awareness, intention and control.* New York: Guilford Press.

Bargh, J.A. (1996). Automaticity in social psychology. In E.T. Higgins & A.W. Kruglanski, *Social psychology: Handbook of basic principles* (pp. 169–183). New York: Guilford Press.

Baron, R.S. (1986). Distraction–conflict theory: Progress and problems. In L. Berkowitz (Ed.), *Advances in Experimental Social Psychology, Vol.19* (pp. 1–40). New York: Academic Press.

Baron, R.S., Moore, D., & Sanders, G.S. (1978). Distraction as a source of drive in social facilitation research. *Journal of Personality and Social Psychology, 36,* 816–824.

Barsalou, L.W. (1985). Ideals, central tendency, and frequency of instantiation. *Journal of Experimental Psychology: Learning, Memory, and Cognition, 11,* 629–654.

Barsalou, L.W. (1988). The content and organization of autobiographical memories. In U. Neisser & E. Winograd (Eds.), *Remembering reconsidered: Ecological and traditional approaches to the study of memory.* Cambridge, UK: Cambridge University Press.

Barsalou, L.W., & Billman, D. (1989). Systematicity and semantic ambiguity. In D. S. Gorfein (Ed.), *Resolving semantic ambiguity.* New York: Springer-Verlag.

Bauer, P. (1993). Memory for gender-consistent and gender-inconsistent event sequences by twenty-five-month-old children. *Child Development, 64,* 285–297.

Baumeister, R.F., Hamilton, J.C., & Tice, D.M. (1985). Public versus private expectancy of success: Confidence booster or performance pressure? *Journal of Personality and Social Psychology, 48,* 1447–1457.

Baumeister, R.F. & Hutton, D.G. (1987). Self-presentation theory: Self-construction and audience pleasing. In B. Mullen & G.R. Goethals (Eds.), *Theories of group behavior* (pp. 71–87). New York: Springer-Verlag.

Baumgardner, A.H., & Brownlee, E.A. (1987). Strategic failure in social interaction: Evidence for expectancy disconfirmation processes. *Journal of Personality and Social Psychology, 52 (3),* 525–535.

Beauvois, J.L., Monteil, J.M., & Trognon, M. (1991). Quelles conduites, quelles cognitions? Repères conceptuels. In J.L. Beauvois, R.V. Joule, & J.M. Monteil (Eds.), *Perspectives Cognitives et Conduites Sociales* (pp. 203–289). Fribourg: DelVal.

Benoit, G., & Everett, J. (1993). Problèmes d'attention et dépression. *L'Année Psychologique, 93,* 401–426.

Besner, D., Stolz, J.A., & Boutilier, C. (1997). The Stroop effect and the myth of automaticity. *Psychonomic Bulletin & Review, 4,* 221–225.

Blank, T.O. (1980). Observer and incentive effects on word association responding. *Personality and Social Psychology Bulletin, 6,* 267–272.

Blank, T.O., Staff, I., & Shaver, P. (1976). Social facilitation of word associations: Further questions. *Journal of Personality and Social Psychology, 34,* 725–733.

Bond, C.F. (1982). Social facilitation: A self-presentational view. *Journal of Personality and Social Psychology, 42,* 1042–1050.

Bond, C.F., & Titus, L.J. (1983). Social facilitation: A meta-analysis of 241 studies. *Psychological Bulletin, 94,* 265–292.

Bower, G.H. (1981). Mood and memory. *American Psychologist, 36,* 129–148.

Bower, G. H. (1991). Mood congruity of social judgements. In J.P. Forgas (Ed.), *Emotion and social judgements* (pp. 31–53). Oxford: Pergamon.

Bower, G.H., & Gilligan, S.G. (1979). Remembering information related to one's self. *Journal of Research in Personality, 13*, 420–432.

Bowers, C.A. (1988). *The cultural dimensions of educational computing: Understanding the non-neutrality of technology.* New York: Teachers College Press.

Brewer, W.F. (1986). What is autobiographical memory? In D.C. Rubin (Ed.), *Autobiographical memory* (pp. 25–49). Cambridge, UK and New York: Cambridge University Press.

Brickman, P., & Bulman, R.J. (1977). Pleasure and pain in social comparison. In J.M. Suls & R.M. Miller (Eds.), *Social comparison processes: Theoretical and empirical perspectives.* Washington, DC: Hemisphere.

Brickner, M., Harkins, S., & Ostrom, T. (1986). Personal involvement. Thought provoking implications for social loafing. *Journal of Personality and Social Psychology, 51*, 763–769.

Broadbent, D.E. (1971). *Decision and stress.* New York: Guilford Press.

Brophy, J.E. (1979). Teacher behavior and its effects. *Journal of Educational Psychology, 71*, 733–750.

Brophy, J.E. (1983). Research on the self-fulfilling prophecy and teacher expectations. *Journal of Educational Psychology, 75*, 631–661.

Brophy, J.E., & Good, T.L. (1970). Teacher's communication of differential expectations for children's classroom performance: Some behavioral data. *Journal of Educational Psychology, 61*, 365–374.

Brown, F.J. (1932). Knowledge of results as an incentive in schoolroom practice. *Journal of Educational Psychology, 23*, 532–552.

Brown, J.D. (1986). Evaluations of self and others: Self-enhancement biases in social judgement. *Social Cognition, 4*, 353–376.

Bruner, J. (1961). The act of discovery. *Harvard Educational Review, 31*, 21–32.

Bruner, J. (1991). *Car la culture donne forme à l'esprit: De la révolution cognitive à la psychologie culturelle.* Paris: Eshel.

Bruning, J.L., Capage, J.E., Kosuh, J.F., Young, P.F., & Young, W.E. (1968). Socially induced drive and range of cue utilization. *Journal of Personality and Social Psychology, 9*, 242–244.

Brunot, S., Huguet, P., & Monteil, J.M. (1999). Performance feedback and self-focused attention in the classroom: When past and present interact. Unpublished Manuscript.

Brunot, S., Monteil, J.M., & Huguet, P. (1995). Performance Feedback and Self-Focused Attention: The Crucial Role of Personal History. *Poster presented at the Joint Meeting of the European Association of Experimental Social Psychology (EAESP) and American Society of Experimental Social Psychology (SESP)*, Washington DC, USA, September.

Burnham, W.H. (1910). The group as a stimulus to mental activity. *Science, 31*, 761–776.

Butler, R. (1990). The effects of mastery and competitive conditions on self-assessment at different ages. *Child Development, 61*, 1420–1433.

Campbell, J.D. (1986). Similarity and uniqueness: The effects of attributes type, relevance, and individual differences in self-esteem and depression. *Journal of Personality and Social Psychology, 50*, 281–294.

Camus, J.F. (1988). La distinction entre les processus controlés et les processus automatiques chez Schneider et Shiffrin. In P. Perruchet (Ed.), *Les automatismes cognitifs.* Liège: Mardaga.

Camus, J.F. (1996). *La psychologie cognitive de l'attention.* Paris: Armand Colin.

Cantor, N., & Mischel, W. (1979). Prototypes in person perception. In L. Berkowitz (Ed.), *Advances in Experimental Social Psychology, Vol. 12.* New York: Academic Press.

Carver, C.S. (1979). A cybernetic model of self-attention processes. *Journal of Personality and Social Psychology, 37*, 1251–1281.

Carver, C.S., & Scheier, M.F. (1978). Self-focusing effects of dispositional self-consciousness, mirror presence, and audience presence. *Journal of Personality and Social Psychology, 36,* 324–332.

Carver, C.S., & Scheier, M.F. (1981a). The self-attention-induced feedback loop and social facilitation. *Journal of Experimental Social Psychology, 17,* 545–568.

Carver, C.S. & Scheier, M.F. (1981b). *Attention and self-regulation: A control theory approach to human behavior.* New York: Springer-Verlag.

Chambon, M. (1990). La perception d'une discipline scolaire par les élèves: Représentation et effets identitaires. *European Journal of Psychology of Education, 6,* 337–354.

Changeux, J.P. (1983). *L'homme neural.* Paris: Fayard.

Charbonnier, E., Huguet, P., Brauer, M., & Monteil, J.M. (1998). Social loafing and self-beliefs: People's collective effort depends on the extent to which they distinguish themselves as better than others. *Social Behavior and Personality, 26,* 329–340.

Chernick, R.S. (1990). Effects of interdependent, coactive, and individualized working conditions on pupil's educational computer program performance. *Journal of Educational Psychology, 82,* 691–695.

Chodorow, N. (1978). *The reproduction of mothering: Psychoanalysis and the sociology of gender.* Berkeley: University of California Press.

Cialdini, R.B., Borden, R.J., Thorne, A., Walker, M.R., Freeman, S., & Sloan, L.R. (1976). Basking in reflected glory: Three field studies. *Journal of Personality and Social Psychology, 34,* 366–375.

Codol, J.P. (1987). Comparability and incomparability between oneself and other: Means of differentiation and comparison reference points. *Cahiers de Psychologie Cognitive, 7,* 87–105.

Collins, R.L. (1996). For better or worse: The impact of upward social comparison on self-evaluations. *Psychological Bulletin, 119,* 51–69.

Comer, D.R. (1995). A model of social loafing in real work group. *Human Relations, 48,* 647–667.

Condry, J. (1977). Enemies of exploration: Self-initiated versus other-initiated learning. *Journal of Personality and Social Psychology, 35,* 459–475.

Conway, M.A. (1987). Verifying autobiographical facts. *Cognition, 26,* 39–58.

Conway, M.A. (1990a). *Autobiographical memory: An introduction.* Buckingham, UK: Open University Press.

Conway, M.A. (1990b). Associations between autobiographical memories and concepts. *Journal of Experimental Psychology: Learning, Memory and Cognition, 16,* 799–812.

Conway, M.A. (1991). In defense of everyday memory. *American Psychologist, 46,* 19–26.

Conway, M.A., & Bekerian, D.A. (1987). Organization in autobiographical memory. *Memory and Cognition, 15,* 119–132.

Conway, M.A., Difazio, R., & Bonneville, J. (1991). Sex, sex roles and response styles for negative affect: Selectivity in a free recall task. *Sex Roles, 25,* 687–700.

Conway, M.A., & Ross, M. (1984). Getting what you want by revising what you had. *Journal of Personality and Social Psychology, 47,* 738–748.

Cottrell, N.B. (1968). Performance in the presence of other human beings: Mere presence, audience, and affiliation effects. In E.C. Simmel, R.A. Hoppe, & G.A. Milton (Eds.), *Social facilitation and imitative behavior* (pp. 91–110). Boston: Allyn & Bacon.

Cottrell, N.B. (1972). Social facilitation. In C.G. McClintock (Ed.), *Experimental social psychology* (pp. 185–236). New York: Holt, Rinehart & Winston.

Cottrell, N.B., Sekerak, G.J., Wack, D.L., & Rittle, R.H. (1968). Social facilitation of dominant responses by the presence of an audience and the mere presence of others. *Journal of Personality and Social Psychology, 9,* 245–250.

Cowan, N. (1988). Evolving conception of memory storage, selective attention, and their mutual constraints within the human information-processing system. *Psychological Bulletin, 104*, 163–191.

Crockenberg, S., & Bryant, B. (1978). Socialization: The "implicit curriculum" of learning environments. *Journal of Research and Development in Education, 12*, 69–78.

Cunningham, M.R., Steinberg, J., & Grev, R. (1980). Wanting to and having to help: Separate motivations for positive and guilt induced helping. *Journal of Personality and Social Psychology, 38*, 181–192.

Dakin, S., & Arrowood, A.J. (1981). The social comparison of ability. *Human Relations, 34*, 89–109.

Damon, W. (1977). *The social world of the child.* San Francisco: Jossey-Bass.

Dashiell, J.F. (1930). An experimental analysis of some group effects. *Journal of Abnormal and Social Psychology, 25*, 190–199.

D'Attore, A. (1981). Computer-aided instruction, boon or bust? *Compute, 12*, 18–20.

Daubman, K.A., Heatherington, L., & Ahn, A. (1992). Gender and the self-presentation of academic achievement. *Sex Roles, 27*, 187–204.

Deci, E.L. (1975). *Intrinsic motivation.* New York: Plenum Press.

Deutsch, M. (1949). An experimental study of the effects of cooperation and competition upon group process. *Human Relations, 2*, 199–232

Doise, W. (1982). *L'explication en psychologie sociale.* Paris: Presses Universitaires de France.

Doise, W. (1989). Constructivism in Social Psychology. *European Journal of Social Psychology, 19*, 389–400.

Doise, W. (1991). Interactions sociales et développement des instruments cognitifs chez l'enfant. In H. Malewska-Peyre & P. Tap (Eds.), *La socialisation de l'enfance à l'adolescence* (pp. 21–47). Paris: Presses Universitaires de France.

Doise, W., & Mugny, G. (1975). Recherches socio-génétiques sur la coordination d'actions interdépendantes. *Revue Suisse de Psychologie Pure et Appliquée, 34*, 160–174.

Doise, W., & Mugny, G. (1981). Le développement social de l'intelligence. Paris: Inter-Editions.

Doise, W., & Mugny, G. (1984). *The social development of the intellect.* Oxford: Pergamon Press.

Dritschel, B.H., Williams, J.M.G., Baddeley, A.D., & Nimmo-Smith, I. (1992). Autobiographical fluency: A method for the study of personal memory. *Memory and Cognition, 20*, 133–140.

Dua, J.K. (1977). Effect of audience on acquisition and extension of avoidance. *British Journal of Social and Clinical Psychology, 16*, 207–212.

Dumas, F., Huguet, P., Dambrun, M., & Despres, G. (1998). Upward comparison leads to better performance: An experiment using the Stroop task. Unpublished Manuscript.

Dumas, F., Huguet, P., & Despres, G. (1998). Comparaison Sociale et Automatismes Cognitifs. Poster presented at the *Deuxième Congrès International de Psychologie Sociale en Langue Française.* Turin, Italy, September.

Duval, S., & Wicklund, R.A. (1972). *A theory of objective self-awareness.* New York: Academic Press.

Easterbrook, J.A. (1959). The effect of emotion on cue utilization and organization of behavior. *Psychological Review, 66*, 183–201.

Eder, D. (1983). Ability grouping and students' academic self-concepts: A case study. *The Elementary School Journal, 84*, 149–161.

Ellis, H.C., & Ashbrook, P.W. (1988). Resource allocation model of the effects of depressed mood states on memory. In K. Fielder & J.P. Forgas (Eds.), *Affect, cognition and social behavior.* Toronto: Hogrefe.

Ellis, H.C., & Ashbrook, P.W. (1989). The "state" of mood and memory research: A selective review. In D. Kuiken (Ed.), *Mood and Memory: Theory, Research, and Applications. Journal of Social Behavior and Personality, 4,* (special issue).

Ellis, H.C., Thomas, R.L., & Rodriguez, I.A. (1984). Emotional mood states and memory: Elaboration encoding, semantic processing, and cognitive effort. *Journal of Experimental Psychology: Learning, Memory, and Cognition, 10,* 470–482.

Fayol, M., & Monteil, J.M. (1988). The notion of "Script": From general to developmental and social psychology. *European Bulletin of Cognitive Psychology, 4,* 335–361.

Fernandez, A., & Glenberg, A.M. (1985). Changing environmental context does not reliably affect memory. *Memory and Cognition, 13,* 333–345.

Festinger, L. (1954). A theory of social comparison processes. *Human Relations, 7,* 117–140.

Fodor, J.A. (1983). *The modularity of mind.* Cambridge, MA: MIT/Bradford Press.

France-Kaatrude, A.C., & Smith, W.P. (1985). Children's interest in social comparison information. *Paper presented at the biennial meeting of the Society for Research in Child Development,* Toronto, Ontario, Canada.

François, S., & Monteil, J.M. (1995). Insertions émotionnelles et performances cognitives. *Communication aux XXV Journées d'études APSLF. Symposium International: La Relation entre Cognition et Emotion,* Coimbra, Portugal.

Freud, S. (1959). The defense neuro-psychoses. In E. Jones & J. Rickman (trans.), *Sigmund Freud: Collected Papers* (Vol. 1). New York: Basic Books (originally published, 1894).

Frey, K.S., & Ruble, D.N. (1985). What children say when the teacher is not around: Conflicting goals in social comparison and performance assessment in the classroom. *Journal of Personality and Social Psychology, 48,* 550–562.

Frey, K.S., & Ruble, D.N. (1990). Strategies for comparative evaluation: Maintaining a sense of competence across the lifespan. In R.J. Sterneberg & J. Kolligian (Eds.), *Perceptions of competence and incompetence across the lifespan* (pp. 167–189). New Haven, CT: Yale University Press.

Gabrenya, W.K., Latané, B., & Wang, Y.E. (1983). Social loafing in cross-cultural perspective: Chinese in Taiwan. *Journal of Cross-Cultural Psychology, 14,* 368–384.

Gabrenya, W.K., Wang, Y.E., & Latané, B. (1985). Social loafing on an optimizing task. Cross cultural differences among Chinese and Americans. *Journal of Cross-Cultural Psychology, 16,* 223–242.

Gastorf, J., & Suls, J.M. (1978). Performance evaluation via social comparison: Performance similarity versus related-attribute similarity. *Journal of Personality and Social Psychology, 41,* 297–305.

Geen, R.G. (1976). Test anxiety, observation, and range of cue utilization. *British Journal of Clinical and Social Psychology, 15,* 253–259.

Geen, R.G. (1979). Effects of being observed on learning following success and failure experiences. *Motivation and Emotion, 3,* 355–370.

Geen, R.G. (1991). Social motivation. *Annual Review of Social Psychology, 42,* 377–399.

Geen, R.G., & Gange, J.J. (1977). Drive theory of social facilitation: Twelve years of theory and research. *Psychological Bulletin, 84,* 1267–1288.

Geller, V., & Shaver, P. (1976). Cognitive consequences of self-awareness. *Journal of Experimental Social Psychology, 12,* 99–108.

Gergen, K.J. (1989). Induction and construction: Teetering between worlds. *European Journal of Social Psychology, 19,* 431–438.

Gibson, J.J. (1979). *The ecological approach to visual perception.* Boston: Houghton Mifflin.

Gilbert, D.T., Giesler, R.B., & Morris, K.A. (1996). When comparisons arise. *Journal of Personality and Social Psychology, 69,* 227–236.

Gilligan, C. (1982). *In a different voice: Psychology theory and women's development.* Cambridge, MA: Harvard University Press.

Gilliland, A.R. (1925). The effect of practice with and without knowledge of results in grading handwriting. *Journal of Educational Psychology, 16*, 532–536.

Glachan, M.D., & Light, P. (1982). Peer interaction and teaching: Can two wrongs make a right? In G. Butterworth & P. Light (Eds.), *Social cognition: Studies of the development of understanding*. Chicago: Chicago University Press.

Glaser, A.N. (1979). The effects of the presence of others: A social-psychological investigation. Unpublished doctoral manuscript.

Glaser, A.N. (1982). Drive theory of social facilitation: A critical reappraisal. *British Journal of Social Psychology, 21*, 265–282.

Goethals, G.R. (1986). Social comparison theory: Lost and found. *Personality and Social Psychology Bulletin, 12*, 261–278.

Goethals, G.R., Messick, D.M., & Allison, S.T. (1991). The uniqueness bias: Studies of constructive social comparison. In J. Suls & T.A. Wills (Eds.), *Social comparison: Theory and research*. Hillsdale, NJ: Lawrence Erlbaum Associates.

Goffman, E. (1959). *The presentation of self in everyday life*. Garden City, New York: Doubleday.

Goffman, E. (1967). *Interaction ritual*. Garden City, New York: Doubleday.

Good, K.J. (1973). Social facilitation: Effects of performance anticipation, evaluation, and response competition on free associations. *Journal of Personality and Social Psychology, 28*, 270–275.

Goodenow, C. (1992). Strengthening the links between educational psychology and the study of social contexts. *Educational Psychologist, 27*, 177–196.

Greenwald, A.G., & Banaji, M.R. (1989). The self as memory system: Powerful, but ordinary. *Journal of Personality and Social Psychology, 57*, 41–54.

Guerin, B. (1986). Mere presence effects in humans: A review. *Journal of Experimental Social Psychology, 22*, 38–77.

Guerin, B. (1993). *Social Facilitation*, European Monographs in Social Psychology. Cambridge, UK: Cambridge University Press.

Halisch, F., & Heckhausen, H. (1977). Search for feedback information and effort regulation during task performance. *Journal of Personality and Social Psychology, 35*, 724–733.

Hampson, P.J. (1989). Aspects of attention and cognitive science. *The Irish Journal of Psychology, 10*, 261–275.

Harkins, S.G. (1987). Social loafing and social facilitation. *Journal of Experimental Social Psychology, 23*, 1–18.

Harkins, S.G., & Jackson, J. (1985). The role of evaluation in eliminating social loafing. *Personality and Social Psychology Bulletin, 11*, 457–465.

Harkins, S.G., & Petty, R. (1982). Effects of task difficulty and task uniqueness on social loafing. *Journal of Personality and Social Psychology, 43*, 1224–1229.

Harkins, S.G., & Szymanski, K. (1988). Social loafing and self-evaluation with an objective standard. *Journal of Experimental Social Psychology, 24*, 354–365.

Harkins, S.G., & Szymanski, K. (1989). Social loafing and group evaluation. *Journal of Personality and Social Psychology, 56*, 934–941.

Harré, R. (1989). Metaphysics and methodology: Some prescriptions for Social Psychology Research. *European Journal of Social Psychology, 19*, 439–454.

Hartwick, J., & Nagao, D.H. (1990). Social facilitation effects in recognition memory. *British Journal of Social Psychology, 29*, 193–210.

Hasher, L., & Zacks, R.T. (1979). Automatic and effortful processes in memory. *Journal of Experimental Psychology: General, 108*, 356–388.

Heider, F. (1958). *The psychology of interpersonal relations*. New York: Wiley.

Henchy, T., & Glass, D.C. (1968). Evaluation apprehension and the social facilitation of dominant and subordinate responses. *Journal of Personality and Social Psychology, 10*, 446–454.

Higgins, E.T. (1987). Self-discrepancy: A theory relating self and affect. *Psychological Review*, *94*, 319–340.

Higgins, E.T. (1989). Continuities and discontinuities in self regulatory and self-evaluative processes: A developmental theory relating self and affect. *Journal of Personality*, *57*, 407–444.

Higgins, E.T., & Bargh, J.A. (1987). Social cognition and social perception. *Annual Review of Psychology*, *38*, 369–425.

Hilton, J.L., Klein, J.G., & Von Hippel, W. (1991). Attention allocation and impression formation. *Personality and Social Psychology Bulletin*, *17*, 548–559.

Hintzman, D.L. (1986). "Schema abstraction" in a multiple-trace memory model. *Psychological Review*, *93*, 411–428.

Hoffman, R.R., & Nead, J.M. (1983). General contextualism, ecological science and cognitive research. *Journal of Mind and Behavior*, *4*, 507–534.

Hokoda, A.J., Fincham, F.D., & Diener, C.I. (1989). The effects of social comparison information on learned helpless and mastery-oriented children in achievement settings. *European Journal of Social Psychology*, *19*, 527–542.

Holland, C.A. (1992). The wider importance of autobiographical memory research. In M. Conway, D.C. Rubin, W.A. Wagenaar, & H. Spinnler (Eds.), *Theoretical perspective on autobiographical memory* (pp. 195–206). Amsterdam: Kluwer.

Huguet, P. (1992). Catégorisations, insertions sociales et performances cognitives: Approche expérimentale. Unpublished Doctoral Dissertation.

Huguet, P. (1993). Social facilitation: A complex problem, not yet resolved. *International Review of Social Psychology*, *6*, 147–151.

Huguet, P. (1995). Travail collectif et performance individuelle. In G. Mugny, S. Oberlé, and J.L. Beauvois (Eds.), *Relations humaines, groupes et influence sociale* (pp. 31–41). Grenoble: Presses Universitaires de Grenoble.

Huguet, P., Chambres, P., & Blaye, A. (1994). Interactive learning: Does social presence explain the results? In H.C. Foot, C.J. Howe, A. Anderson, A.K. Tolmie, & D.A. Warden (Eds.), *Group and interactive learning* (pp. 437–443). Southampton: Computational Mechanics Publications.

Huguet, P., Charbonnier, E., & Monteil, J.M. (1995a). Where does self-esteem come from? The influence of private versus public individuation. *Social Behavior and Personality*, *23*, 70–81.

Huguet, P., Charbonnier, E., & Monteil, J.M. (1998a). Social loafing: People's collective effort depends on the extent to which they distinguish themselves as better than others. Unpublished Manuscript.

Huguet, P., Charbonnier, E., & Monteil, J.M. (in press, a). Productivity loss in performance groups: People who see themselves as average do not engage in social loafing. *Group Dynamics: Theory, Research, and Practice*.

Huguet, P., Galvaing, M.P., Dumas, F., & Monteil, J.M. (in press, b). The social influence of automatic responding: Controlling the uncontrollable. In J.P. Forgas, K.D. Williams, & L. Wheeler (Eds.), *The Social Mind: Cognitive and motivational aspects of interpersonal behaviour*. Cambridge: Cambridge University Press.

Huguet, P., Galvaing, M.P., Michinov, N., & Monteil, J.M. (1997). Audience, coaction, and social comparison: Their effects on Stroop interference. *Paper presented at the Nags Head Conference on Social Comparison*, Highland Beach, Florida, USA, June.

Huguet, P., & Latané, B. (1996). Social representations as dynamic social impact. *Journal of Communication*, *46*, 57–63.

Huguet, P., Latané, B., & Bourgeois, M. (1998b). The emergence of a social representation of human rights: Empirical evidence for the convergence of two theories. *European Journal of Social Psychology*, *28*, 831–846.

Huguet, P., & Monteil, J.M. (1992). Social comparison and cognitive performance: A descriptive approach in an academic context. *European Journal of Psychology of Education, 7,* 131–150.

Huguet, P., & Monteil, J.M. (1994). Learning in co-acting groups: The role of social comparison and gender norms. In H.C. Foot, C.J. Howe, A. Anderson, A.K. Tolmie, & D.A. Warden (Eds.), *Group and interactive learning* (pp. 445–451). Southampton: Computational Mechanics Publications.

Huguet, P., & Monteil, J.M. (1995). The influence of social comparison with less fortunate others on task performance: The role of gender motivations or appropriate norms. *Sex Roles, 33,* 753–765.

Huguet, P., & Monteil, J.M. (1996). Comparaison sociale personnelle, genre et performances cognitives: Etude expérimentale de l'intervention des appartenances de sexe dans le fonctionnement cognitif individuel. In J.L. Beauvois., R.V. Joule, & J.M. Monteil (Eds.), *Perspectives cognitives et conduites sociales, Vol. 5* (pp. 33–48). Neuchâtel: Delachaux & Niestlé.

Huguet, P., Monteil, J.M., & Charbonnier, E. (1995b). Perceived self-uniqueness as a moderator of social loafing. *Poster presented at the Joint Meeting of the European Association of Experimental Social Psychology (EAESP) and American Society of Experimental Social Psychology (SESP),* Washington DC, USA, September.

Huguet, P., Monteil, J.M., & Charbonnier, E. (1996a). Self-uniqueness moderates social loafing. *Paper presented at the Nags Head Conference on Groups, Networks, and Organizations,* Highland Beach, Florida, USA, June.

Huguet, P., Monteil, J.M., & Galvaing, M.P. (1996b). Does social presence increase or decrease the Stroop color-word interference? Empirical evidence for an attentional view of social facilitation. *Paper presented at the 11th General Meeting of the European Association of Experimental Social Psychology (EAESP),* Gmünden, Austria, July.

Huguet, P., Monteil, J.M., Galvaing, M.P., & Charbonnier, E. (in press, c). Présence d'Autrui et Performance Individuelle: Repères et Eléments de Réflexion. *Connexions.*

Huguet, P., Monteil, J.M., Galvaing, M.P., & Dumas, F. (1998c). Social presence effects in the Stroop task: Further evidence for an attentional view of social facilitation. Unpublished manuscript.

Hull, C.L. (1943). *Principles of behavior.* New York: Appleton Century.

Humphrey, N. (1986). *The inner eye.* London: Faber & Faber.

Ingham, A., Levinger, G., Graves, J., & Peckham, V. (1974). The Ringelmann effect: Studies of group size and group performance. *Journal of Experimental Social Psychology, 10,* 371–384.

Innes, J.M., & Young, R.F. (1975). The effect of presence of an audience, evaluation apprehension, and objective self-awareness on learning. *Journal of Experimental Social Psychology, 11,* 35–42.

Israël, J., & Tajfel, H. (1972). *The context of social psychology: A critical assessment.* Academic Press: London.

Jackson, J.M., & Harkins, S.G. (1985). Equity in effort: An explanation of the social loafing effect. *Journal of Personality and Social Psychology, 49,* 1199–1206.

Jackson, J.M., & Williams, K.D. (1985). Social loafing on difficult tasks: Working collectively can improve performance. *Journal of Personality and Social Psychology, 49,* 937–942.

Jacoby, L.L., & Dallas, M. (1981). On the relationship between autobiographical memory and perceptual learning. *Journal of Experimental Psychology: General, 110,* 306–340.

Johnson, D.W., & Johnson, R.T. (1974). Instructional goal structure: Cooperative, competitive, or individualistic. *Review of Educational Research, 44,* 213–224.

Johnson, D.W., & Johnson, R.T. (1989). *Cooperation and competition: Theory and research.* Edina, MN: Interaction Book Company.

Johnson, D.W., & Johnson, R.T. (1991). *Learning together and alone: Cooperative, competitive and individualistic learning.* Boston: Allyn & Bacon.

Johnson, D.W., Johnson, R.T., & Scott, L. (1978). The effects of cooperative and individualized instruction on student attitudes and achievement. *Journal of Social Psychology, 104,* 207–216.

Johnston, W.A., & Dark, V.J. (1986). Selective attention. *Annual Review of Psychology, 37,* 43–75.

Josephs, R.A., Markus, H.R., & Tafarodi, R.W. (1992). Gender and self-esteem. *Journal of Personality and Social Psychology, 63,* 391–402.

Kagan, D. (1990). How schools alienate students at risk: A model for examining proximal classroom variables. *Educational Psychologist, 25,* 105–126.

Kahneman, D. (1973). *Attention and effort.* Englewood Cliffs, NJ: Prentice-Hall.

Kantowitz, B.H. (1985). Channels and stage in human information processing: A limited analysis of theory and methodology. *Journal of Mathematical Psychology, 29,* 135–174.

Karau, S.J., & Williams, K.D. (1993). Social loafing: A meta-analytic review and theoretical integration. *Journal of Personality and Social Psychology, 65,* 681–706.

Kashima,Y., Yamagishi, S., Kim, U., Choi, S.C., Gelfand, M.J., & Yuki, M. (1995). Culture, gender, and self: A perspective from individualism–collectivism research. *Journal of Personality and Social Psychology, 69,* 925–937.

Kerr, N. (1983). Motivation losses in small groups: A social dilemma analysis. *Journal of Personality and Social Psychology, 45,* 819–828.

Kerr, N., & Bruun, S. (1981). Ringelmann revisited: Alternative explanations for the social loafing effect. *Personality and Social Psychology Bulletin, 7,* 224–231.

Kerr, N. & Bruun, S. (1983). The dispensability of member effort and group motivation losses: Free rider effects. *Journal of Personality and Social Psychology, 44,* 78–94.

Kihlström, J.F., Cantor, N., Albright, J.S., Chew, B.R., Klein, S.B., & Niedenthal, P.M. (1988). Information processing and the study of the self. *Advances in Experimental Social Psychology, 21,* 145–178.

Kluger, A.N., & DeNisi, A. (1996). The effects of feedback interventions on performance: A historical review, a meta-analysis, and a preliminary feedback intervention theory. *Psychological Bulletin, 119,* 254–284.

Kravitz, D., & Martin, B. (1986). Ringelmann rediscovered: The original article. *Journal of Personality and Social Psychology, 50,* 936–941.

Kruglanski, A.W., & Mayseless, O. (1990). Classic and current social comparison research: Expanding the perspective. *Psychological Bulletin, 108,* 195–208.

Kuiper, N.A., & Derry, P.A. (1981). The self as a cognitive prototype: An application to person perception and depression. In N. Cantor & J.F. Kihlström (Eds.), *Personality, cognition and social interaction* (pp. 215–232). Hillsdale, NJ: Erlbaum.

Kushnir, T. (1981). The status of arousal in recent social facilitation literature: A review and evaluation of assumptions implied by the current research model. *Social Behaviour and Personality, 9,* 185–190.

Kushnir, T., & Duncan, K.D. (1978). An analysis of social facilitation effects in terms of signal detection theory. *Psychological Record, 28,* 535–541.

Larsen, S.F. (1985). Specific background knowledge and updating. In J. Allwood & E. Hjelmquist (Eds.), *Foregrounding background.* Lund: Doxa.

Larsen, S.F (1992). Personal context in autobiographical and narrative memories. In M. Conway, D.C. Rubin, W.A. Wagenaar, & H. Spinnler (Eds.), *Theoretical perspective on autobiographical memory* (pp. 53–71). Amsterdam: Kluwer.

Latané, B. (1981). The psychology of social impact. *American Psychologist, 36,* 343–356.

Latané, B. (1996). Dynamic social impact theory: The creation of culture by communication. *Journal of Communication, 46,* 13–25.

Latané, B., Nowak, A., & Liu, J.H. (1994). Measuring emergent social phenomena: Dynamism, polarization, and clustering as order parameters of social systems. *Behavioral Science, 39,* 1–24.

Latané, B., Williams, K., & Harkins, S. (1979). Many hands make light the work: The causes and consequences of social loafing. *Journal of Personality and Social Psychology, 37,* 822–832.

Latham, G.P., & Locke, E.A. (1991). Self-regulation through goal setting. *Organizational Behavior and Human Decision Processes, 50,* 212–247.

Lecourt, D. (1989). Introduction. In A. Prochiantz (Ed.), *La construction du cerveau* (pp. 1–17). Paris: Hachette.

Lemaine, G. (1966). Inegalité, comparaison et incomparabilité: Esquisse d'une théorie de l'originalité sociale. *Bulletin de Psychologie, 20,* 24–32.

Lemaine, G. (1974). Social differenciation and social originality. *European Journal of Social Psychology, 4,* 17–52.

Lepper, M.R., & Greene, D. (1978). *The hidden costs of reward.* Hillsdale, NJ: Erlbaum.

Lepper, M.R., & Hodell, M. (1989). Intrinsic motivation in the classroom. In C. Ames & R. Ames (Eds.), *Research on motivation in education* (pp. 73–105). San Diego, CA: Academic Press.

Levine, J.M. (1983). Social comparison and education. In J.M. Levine & M.C. Wang (Eds.), *Teacher and student perceptions: Implications for learning.* Hillsdale, NJ: Erlbaum.

Lewicki, P. (1983). Self-image bias in person perception. *Journal of Personality and Social Psychology, 45,* 384–393.

Leyens, J.P. (1983). *Sommes-nous tous des psychologues?* Brussels: Mardaga.

Leyens, J.P., Yzerbyt, V.Y., & Schadron, G. (1994). *Stereotypes and social cognition.* London: Sage.

Logan, G.D. (1978). Attention in character-classification tasks: Evidence for the automaticity of component stages. *Journal of Experimental Psychology: General, 107,* 32–63.

Logan, G.D. (1979). On the use of a concurrent memory load to measure attention and automaticity. *Journal of Experimental Psychology: Human Perception and Performance, 5,* 189–207.

Logan, G.D. (1980). Attention and automaticity in Stroop and priming tasks: Theory and data. *Cognitive Psychology, 12,* 523–553.

Logan, G.D. (1988). Toward an instance theory of automatization. *Psychological Review, 95,* 492–527.

Lorenzi-Cioldi, F. (1988). *Individus dominants et groupes dominés.* Grenoble: Presses Universitaires de Grenoble.

MacLeod, C.M. (1991). Half a century of research on the Stroop effect: An integrative review. *Psychological Bulletin, 109,* 163–203.

Manstead, A.S.R., & Semin, G.R. (1980). Social facilitation effects: Mere enhancement of dominant response? *British Journal of Social and Clinical Psychology, 19,* 119–136.

Manzer, C.W. (1935). The effect of knowledge of output on muscular work. *Journal of Experimental Psychology, 8,* 80–90.

Maracek, J., & Mettee, D.R. (1972). Avoidance of continued success as a function of self-esteem, level of esteem certainty and responsibility for success. *Journal of Personality and Social Psychology, 22,* 98–107.

Marks, G. (1984). Thinking one's abilities are unique and one's opinions are common. *Personality and Social Psychology Bulletin, 10,* 203–208.

Markus, H.R. (1977). Self-schemata and processing information about the self. *Journal of Personality and Social Psychology, 35,* 63–78.

Markus, H.R., & Kitayama, S. (1991). Culture and the self: Implications for cognition, emotion, and motivation. *Psychological Review, 98,* 224–253.

Markus, H.R., & Smith, J. (1981). The influence of self-schemata on the perception of others. In N. Cantor & J.F. Kihlström (Eds.), *Personality, cognition, and social interaction* (pp. 233–262). Hillsdale, NJ: Erlbaum.

Markus, H.R., & Wurf, E. (1987). The dynamic self-concept: A social psychological perspective. *Annual Review of Psychology, 38,* 299–337.

Marsh, H.W. (1990). Influences of internal and external frames of reference on the formation of math and english self-concepts. *Journal of Educational Psychology, 82,* 107–116.

Marsh, H. W., & Parker, J. (1984). Determinants of student self-concept: Is it better to be a relatively large fish in a small pond even if you don't learn to swim as well? *Journal of Personality and Social Psychology, 47,* 213–231.

Marshall, H.H., & Weinstein, R.S. (1984). Classroom factors affecting students' self-evaluations: An interactional model. *Review of Educational Research, 54,* 301–325.

Martin, L.L., Seta, J.J., & Crelia, R.A. (1990). Assimilation and contrast as a function of people's willingness and ability to expend effort in forming an impression. *Journal of Personality and Social Psychology, 59,* 27–37.

Martinot, D. (1993). *Traitement de l'information autopertinente et information sociale: Le soi, une structure de connaissances malléable. Etudes expérimentales.* Unpublished Doctoral Dissertation.

Martinot, D. (1995). *Le Soi: Les approches psychosociales.* Grenoble: Presses Universitaires de Grenoble.

Martinot, D., & Monteil, J.M. (1995). Academic self-schema: Toward an experimental illustration. *Learning and Instruction, 5,* 63–76.

Martinot, D., & Monteil, J.M. (1996). Insertions sociales et schémas de soi scolaires de réussite et d'échec: Etude expérimentale. In J.L. Beauvois., R.V. Joule, & J.M. Monteil (Eds.), *Perspectives cognitives et conduites sociales (Vol. 5).* Neuchâtel: Delachaux & Niestlé.

Mash, E.J., & Hedley, J. (1975). Effect of observer as a function of prior history of social interaction. *Perceptual and Motor Skills, 40,* 659–669.

Masters, J.C. (1971). Effects of social comparison upon children's self-reinforcement and altruism toward competitors and friends. *Developmental Psychology, 5,* 64–72.

Masters, J.C., Carlston, C.R., & Raye, D.F. (1985). Children's affective, behavioral, and cognitive responses to social comparison. *Journal of Experimental Social Psychology, 21,* 407–420.

Masters, J.C., & Keil, L.J. (1987). Generic comparison processes in human judgement and behavior. In J.C. Masters & W.P. Smith (Eds.), *Social comparison, social justice, and relative deprivation: Theoretical, empirical, and policy perspectives* (pp. 11–54). Hillsdale, NJ: Erlbaum.

Mead, G.H. (1934). *Mind, self, and society.* Chicago: Chicago University Press.

Mead, G.H. (1963). *L'esprit, le soi et la société.* Paris: Presses Universitaires de France.

Mettee, D.R. (1971). Rejection of unexpected success as a function of the negative consequences of accepting success. *Journal of Personality and Social Psychology, 17,* 332–341.

Michinov, N. (1997). *Etudes expérimentales de quelques déterminants des stratégies de comparaison sociale* (Doctoral Dissertation). Villeneuve d'Ascq: Presses Universitaires du Septentrion.

Miller, C.T. (1984). Self-schemas, gender, and social comparison: A clarification of the related attributes hypothesis. *Journal of Personality and Social Psychology, 46,* 1222–1229.

Miller, G.A., Galanter, E., & Pribam, K.H. (1960). *Plans and the structure of behavior.* New York: Holt, Rinehart, & Winston.

Miller, J.B. (1986). *Toward a new psychology of women* (2nd ed.). Boston: Beacon Press.

Minsky, M. (1985). *The society of mind.* New York: Simon & Schuster.

Monteil, J.M. (1988). Comparaison sociale. Stratégies individuelles et médiations socio-cognitives. Un effet de différenciations comportementales dans le champ scolaire. *European Journal of Psychology of Education, 3,* 3–18.

Monteil, J.M. (1989). *Eduquer et former: Perspectives psychosociales.* Grenoble: Presses Universitaires de Grenoble.

Monteil, J.M. (1991). Social regulations and individual cognitive functioning: Effects of individuation on cognitive performances. *European Journal of Social Psychology, 21,* 255–277.

Monteil, J.M. (1992a). Toward a social psychology of cognitive functioning: Theoretical outline and empirical illustrations. In W. Doise & G. Mugny (Eds.), *Social representations and the social bases of knowledge* (pp. 45–55). Berne: Hubert.

Monteil, J.M. (1992b). Intergroup differentiation and individualization: The effect of social deprivation. *European Bulletin of Cognitive Psychology, 12,* 189–203.

Monteil, J.M. (1993a). *Soi et le contexte.* Paris: Armand Colin.

Monteil, J.M. (1993b). Towards a social psychology of cognition. In M.C. Hurtig, M.F. Pichevin, & M. Piolat (Eds.), *Self-concept and social cognition* (pp. 209–219). Singapore: Scientific Writings.

Monteil, J.M. (1994). Interactions sociales. In M. Richelle, M. Robert, & J. Requin (Eds.), *Traité de Psychologie Expérimentale* (pp. 133–179). Paris: Presses Universitaires de France.

Monteil, J.M. (1995). Insertions sociales et performances cognitives. *Psychologie Française, 40,* 319–329.

Monteil, J.M. (1997). Contexte social et performances scolaires. Préliminaires pour une théorie du feedback de comparaison sociale. In J.L. Beauvois, R.V. Joule, & J.M. Monteil (Eds.), *15 ans de recherches expérimentales en psychologie sociale francophone.* Paris: Dunod.

Monteil, J.M., Bavent, L., & Lacassagne, M.F. (1986). Attribution et mobilisation d'une appartenance idéologique: Un effet polydoxique. *Psychologie Française, 31,* 115–121.

Monteil, J.M., Brunot, S., & Huguet, P. (1996). Cognitive performance and attention in the classroom: An interaction between past and present academic experiences. *Journal of Educational Psychology, 88,* 242–248.

Monteil, J.M., & Castel P. (1989). Modes d'insertions sociaux, attribution de sanctions et comparaisons sociales: Une contribution expérimentale à l'explication de conduites scolaires. In J.L. Beauvois, R.V. Joule., & J.M. Monteil (Eds.), *Perspectives cognitives et conduites sociales,* vol. 2, (pp. 299–310). Fribourg: Delval.

Monteil, J.M., & Chambres, P. (1990). Eléments pour une exploration des dimensions du conflit socio-cognitif: Une expérimentation chez l'adulte. *Revue Internationale de Psychologie Sociale, 3,* 499–518.

Monteil, J.M., & Huguet, P. (1991). Insertion sociale, catégorisation sociale et activités cognitives. *Psychologie Française, 36,* 35–46.

Monteil, J.M., & Huguet, P. (1993a). The influence of social comparison situations on individual task performance: Experimental illustrations. *International Journal of Psychology, 28,* 627–634.

Monteil, J.M., & Huguet, P. (1993b). The social context of human learning: Some prospects for the study of socio-cognitive regulations. *European Journal of Psychology of Education, 8,* 409–421.

Monteil, J.M., & Martinot, D. (1991). Le soi et ses propriétés: Analyse critique. *Psychologie Française, 36,* 55–66.

Monteil, J.M., & Michinov, N. (1996). Study of some determinants of social comparison strategies by a new methodological tool. Toward a dynamical approach. *European Journal of Social Psychology, 26,* 801–819.

Monteil, J.M., & Michinov, N. (1998). Effects of context and performance feedback on social comparison strategies among low-achievement students: An experimental study. Unpublished Manuscript.

Moore, D.L., & Baron, R.S. (1983). Social facilitation: A psychophysiological analysis. In J. Cacioppo & R. Petty (Eds.), *Social psychophysiology: A sourcebook* (pp. 434–466). New York: Guilford Press.

Moray, N. (1962). Review of annual review of psychology. *Quarterly Journal of Experimental Psychology, 14*, 58–60.

Morse, S., & Gergen, K.J. (1970). Social comparison, self-consistency, and the concept of self. *Journal of Personality and Social Psychology, 16*, 148–156.

Mosatche, H.S., & Bragonnier, P. (1981). An observational study of social comparison in preschoolers. *Child Development, 52*, 376–378.

Moscovici, S. (1961). *La psychanalyse: Son image et son public*. Paris: Presses Universitaires de France.

Moscovici, S. (1984). The phenomenon of social representations. In R.M. Farr & S. Moscovici (Eds.), *Social representations* (pp. 3–69). Cambridge, UK: Cambridge University Press.

Moscovici, S. (1989). Preconditions for explanations in social psychology. *European Journal of Social Psychology, 19*, 407–430.

Mugny, G., & Carugati, F. (1985). *L'intelligence au pluriel, les représentations sociales de l'intelligence et de son développement*. Fribourg: DelVal.

Mullen, B. (1985). Strength and immediacy of sources: A meta-analytic evaluation of the forgotten elements of social impact theory. *Journal of Personality and Social Psychology, 49*, 1458–1466.

Nakamura, C.Y., & Finck, D. (1973). Effects of social or task orientation and evaluative or nonevaluative situations on performance. *Child Development, 44*, 83–93.

Navon, D., & Gopher, D. (1979). On the economy of the human processing system. *Psychological Review, 86*, 214–255.

Neely, J.H. (1977). Semantic priming and retrieval from lexical memory: Roles of inhibitionless spreading activation and limited-capacity attention. *Journal of Experimental Psychology: General, 106*, 226–254.

Neisser, U. (1986). Nested structure in autobiographical memory. In D.C. Rubin (Ed.), *Autobiographical memory*. Cambridge, UK: Cambridge University Press.

Neisser, U. (1988). What is ordinary memory the memory of? In U. Neisser & E. Winograd (Eds.), *Remembering reconsidered: Ecological and traditional approach to memory*. Cambridge, UK: Cambridge University Press.

Nelson, K. (1986). *Event knowledge*. Hillsdale, NJ: Erlbaum.

Nicholls, J.G. (1984). Conceptions of ability and achievement motivation: A theory and its implications for education. In S.G. Paris, G.H. Olson, & H.W. Stevenson (Eds.), *Learning and motivation in the classroom*. Hillsdale, NJ: Erlbaum.

Nielsen, S.L., & Sarason, J.G. (1981). Emotion, personality, and selective attention. *Journal of Personality and Social Psychology, 41*, 495–560.

Norman, D.A., & Bobrow, D.G. (1975). On data-limited and resources-limited processes. *Cognitive Psychology, 7*, 44–64.

Nuttin, J.M. (1989). Proposal for a heuristic quasi-social analysis of social behavior: The case of Harlow's "Nature of Love". *European Journal of Social Psychology, 19*, 371–384.

Olson, M. (1987). *Logique de l'action collective*. Paris: Presses Universitaires de France (2nd ed.). (Originally published in 1966 as *Logic of collective action: Public goods and the theory of groups*. Cambridge, MA: Harvard University Press.)

Parasuraman, R., & Davies, D.R. (Eds.) (1985). *Varieties of attention*. Orlando, FL: Academic Press.

Parker, L.E., & Lepper, M.R. (1992). Effects of fantasy contexts on children's learning and motivation: Making learning more fun. *Journal of Personality and Social Psychology, 62,* 625–633.

Paulhus, D.L., Graf, P., & Van Selst, M. (1989). Attentional load increases the positivity of self-presentation. *Social Cognition, 7,* 389–400.

Paulhus, D.L., & Levitt, K. (1987). Desirable responding triggered by affect: Automatic egotism? *Journal of Personality and Social Psychology, 52,* 245–259.

Paulus, P.B., & Murdoch, P. (1971). Anticipated evaluation and audience presence in the enhancement of dominant responses. *Journal of Experimental Social Psychology, 7,* 280–291.

Pepitone, E. (1972). Comparison behavior in elementary school children. *American Journal of Educational Research, 9,* 45–63.

Pepitone, E. (1982). Social comparison and pupil interaction: Effect of homogeneous vs. heterogeneous classrooms. *Communication at the Annual Meeting of the American Educational Research Association,* New York.

Pessin, J. (1933). The comparative effects of social and mechanical stimulation on memorizing. *American Journal of Psychology, 45,* 263–270.

Pettigrew, T.F. (1967). Social evaluation theory: Convergences and applications. In D. Levine (Ed.), *Nebraska Symposium on Motivation* (pp. 241–315). Lincoln, NB: University of Nebraska Press.

Petty, R.E., Harkins, S.G., & Williams, K.D. (1980). The effects of diffusion of cognitive effort on attitudes: An information processing view. *Journal of Personality and Social Psychology, 38,* 81–92.

Piaget, J. (1928). Les trois systèmes de la pensée de l'enfant. *Bulletin de Sociologie Française et de Philosophie, 28,* 97–141.

Posner, M.L., & Snyder, C.R. (1975). Attention and cognitive control. In R.L. Solso (Ed.), *Information processing and cognition,* The Loyola Symposium. Hillsdale, NJ: Erlbaum.

Pritchard, R.D., Jones, S.D., Roth, P.L., Stuebing, K.K., & Ekeberg, S.E. (1988). Effects of group feedback, goal setting, and incentives on organizational productivity. *Journal of Applied Psychology, 73,* 337–358.

Prochiantz, A. (1989). *La construction du cerveau.* Paris: Hachette.

Psotka, J. (1982). Computers and education. *Behavior Research Methods & Instrumentation, 14,* 221–223.

Quine, W.V.O. (1985). Events and reification. In E. LePore & P.P. McLaughlin (Eds.), *Actions and events: Perspectives on the philosophy of Donald Davidson.* Oxford: Blackwell.

Reiser, B.J., Black, J.B., & Abelson, R.P. (1985). Knowledge structures in the organization and retrieval of autobiographical memories. *Cognitive Psychology, 17,* 89–137.

Reiser, B.J., Black, J.B., & Kalamarides, P. (1986). Strategic memory search processes. In D.C. Rubin (Ed.), *Autobiographical memory.* Cambridge, UK and New York: Cambridge University Press.

Richard, J.F. (1980). *L'attention.* Paris: Presses Universitaires de France.

Richard, J.F. (1982). Planification et organisation des actions dans la résolution du problème de la tour de Hanoï par des enfants de 7 ans. *Année Psychologique, 82,* 307–336.

Rijsman, J. (1974). Factors in social comparison of performance influencing actual performance. *European Journal of Social Psychology, 4,* 279–311.

Rijsman, J. (1983). The dynamics of social competition in personal and categorical comparison-situations. In W. Doise & S. Moscovici (Eds.), *Current Issues in European Social Psychology, Vol. 1* (pp. 279–312).

Ringelmann, M. (1913). Recherches sur les moteurs animés: Travail de l'homme. *Annales de l'Institut National Agronomique, 12,* 1–40.

Robinson, J.A. (1986). Autobiographical memory: A historical prologue. In D.C. Rubin (Ed.), *Autobiographical memory*. Cambridge, UK and New York: Cambridge University Press.

Robinson, J.A., & Swanson, K.L. (1989). Is memory perspective related to memory age in autobiographical recall? *Paper presented at the 30th Annual Meeting of the Psychonomic Society*, Atlanta.

Robinson, J.A., & Swanson, K.L. (1990). Autobiographical memory. The next phase. *Applied Cognitive Psychology, 4*, 321–335.

Robinson, W.P. (1984). Social psychology in classrooms. In G.M. Stephenson & J.H. Davis (Eds.), *Progress in applied social psychology* (pp. 93–125). Chichester: Wiley.

Robinson, W.P., & Tayler, C.A. (1989). Correlates of low academic attainment in three countries. *International Journal of Educational Research, 13*, 581–596.

Robinson, W.P., Tayler, C.A., & Piolat, M. (1990). School attainment, self-esteem, and identity: France and England. *European Journal of Social Psychology, 20*, 387–403.

Rogers, T.B. (1981). A model of the self as an aspect of the human information processing system. In N. Cantor & J.F. Kihlström (Eds.), *Personality, cognition and social interaction*. Hillsdale, NJ.: Erlbaum.

Rosenholtz, S.R., & Wilson, B. (1980). The effect of classroom structure on shared perceptions of ability. *American Educational Research Journal, 17*, 75–82.

Rosenthal, R., & Jacobson, L. (1968). *Pygmalion in the classroom. Teacher expectations and pupils' intellectual development*. New York: Holt, Rinehart, & Winston.

Rosenzweig, S. (1933). The experimental situation as a psychological problem. *Psychological Review, 40*, 337–354.

Ross, L., Greene, D., & House, P. (1977). The "false consensus effect": An egocentric bias in social perception and attribution processes. *Journal of Personality and Social Psychology, 13*, 279–301.

Rubin, D.C. (1982). The retention function for autobiographical memory. *Journal of Verbal Learning and Verbal Behavior, 21*, 21–38.

Rubin, D.C. (1986). *Autobiographical memory*. Cambridge, UK and New York: Cambridge University Press.

Ruble, D.N. (1983). The development of social comparison processes and their role in achievement-related self-socialization. In E.T. Higgins, D.N. Ruble, & W.W. Hartup (Eds.), *Social cognition and social development: A sociocultural perspective* (pp. 134–157). New York: Cambridge University Press.

Ruble, D.N., Boggiano, A.K., Feldman, N.S., & Loeble, J.H. (1980). Developmental analysis of the role of social comparison in self-evaluation. *Developmental Psychology, 16*, 105–115.

Ruble, D.N., & Frey, K.S. (1987). Social comparison and self-evaluation in the classroom: Developmental changes in knowledge and function. In J.C. Masters & W.P. Smith (Eds.), *Social comparison, social justice, and relative deprivation* (pp. 81–105). Hillsdale, NJ: Lawrence Erlbaum Associates.

Sanders, G.S. (1981). Driven by distraction: An integrative review of social facilitation theory and research. *Journal of Experimental Social Psychology, 17*, 227–257.

Sanders, G.S. (1984). Self-presentation and drive in social facilitation. *Journal of Experimental Social Psychology, 20*, 312–322.

Sanders, G.S., & Baron, R.S. (1975). The motivating effects of distraction on task performance. *Journal of Personality and Social Psychology, 32*, 956–963.

Sanna, L.J. (1992). Self-efficacy theory: Implications for social facilitation and social loafing. *Journal of Personality and Social Psychology, 62*, 774–786.

Sanna, L.J., & Shotland, R.L. (1990). Valence of anticipated evaluation and social facilitation. *Journal of Experimental Social Psychology, 26*, 82–92.

Santrock, J.W., & Ross, M. (1975). Effects of social comparison on facilitative self-control in young children. *Journal of Educational Psychology, 67*, 193–197.

Santrock, J.W., Smith, P.C., & Bourbeau, P.E. (1976). Effects of social comparison on aggression and regression in groups of young children. *Child Development, 47*, 831–837.

Saufley, W.H. Jr., Otaka, S.R., & Bavaresco, J.L. (1985). Context effects: Classroom tests and context independence. *Memory and Cognition, 13*, 522–528.

Schank, R.C., & Abelson, R.P. (1977). *Scripts, plans, goals and understanding.* Hillsdale, NJ: Erlbaum.

Scheier, M.F., & Carver, C.S. (1988). A model of behavioral self-regulation: Translating intention into action. In L. Berkowitz (Ed.), *Advances in experimental social psychology, 21.* New York: Academic Press.

Schlenker, B.R. (1980). *Impression management: The self-concept, social identity, and interpersonal relations.* Monterey, CA: Brooks/Cole.

Schneider, D.J. (1991). Social cognition. *Annual Review of Psychology, 42*, 527–561.

Schneider, W., & Fisk, A.D. (1982). Degree of consistent training: Improvements in search performance and automatic process development. *Perception and Psychophysics, 31*, 160–168.

Schneider, W., & Shiffrin, R.M. (1977). Controlled and automatic human information processing I: Detection, search, and attention. *Psychological Review, 84*, 1–66.

Searle, J. (1985). *Du cerveau au savoir.* Paris: Hermann.

Sedikides, C. (1992). Mood as a determinant of attentional focus. *Cognition and Emotion, 6*, 129–148.

Seibert, P.S., & Ellis, H.C. (1991). Irrelevant thoughts, emotional mood states and cognitive task performance. *Memory and Cognition, 19*, 507–553.

Seta, J.J., & Hassan, R.K. (1980). Awareness of prior success and failure: A critical factor in task performance. *Journal of Personality and Social Psychology, 39*, 70–76.

Shaver, P., & Liebling, B. (1976). Explorations in the drive theory of social facilitation. *Journal of Social Psychology, 99*, 259–271.

Shepperd, J.A. (1993). Productivity loss in performance groups: A motivation analysis. *Psychological Bulletin, 113*, 67–81.

Sherman, S.J., Judd, C.M., & Park, B. (1989). Social cognition. In M.R. Rosenzweig & L.W. Porter (Eds.), *Annual review of psychology*, Vol. 36. Palo Alto, CA: Annual Review.

Shiffrin, R.M., & Schneider, W. (1977). Controlled and automatic human information processing II: Perceptual learning, automatic attending and a general theory. *Psychology Review, 84*, 127–187.

Simon, H.A. (1976). Discussion: Cognition and social behavior. In J. Carroll & J. Payne (Eds.), *Cognition and social behavior.* Hillsdale, NJ: Erlbaum.

Simon, H.A. (1982). Unity of the arts and science: The psychology of thought and discovery. *American Academy of Arts and Sciences Bulletin, 35*, 6.

Simon, H.A. (1990). Invariants of human behavior. *Annual Review of Psychology, 41*, 1–19.

Singer, J.E. (1966). Social comparison. Progress and issues. *Journal of Experimental Social Psychology, 2 (suppl. 1)*, 103–110.

Singer, J.L. (1977). Imagination and make-believe play in early childhood: Some educational implications. *Journal of Mental Imagery, 1*, 127–144.

Skinner, B.F. (1981). Selection by consequences. *Science, 213*, 501–504.

Slavin, R.E. (1983). *Cooperative learning.* New York: Longman.

Slavin, R.E. (1985). Cooperative learning : Applying contact theory in desegregated schools. *Journal of Social Issues, 41*, 45–62.

Slavin, R.E. (1990). *Cooperative learning. Theory research and practice.* Needham Heights: Allyn and Bacon.

Smith, A.P. (1991). Noise and aspect of attention. *British Journal of Psychology, 80*, 313–324.

Smith, E.R. (1990). Content and process specificity in the effects of prior experiences. In T.K. Srull & R.S. Wyer (Eds.), *Advances in social cognition*, Vol. 3. Hillsdale, NJ: Lawrence Erlbaum Associates.

Smith, R.H., & Insko, C.A. (1987). Social comparison choice during ability evaluation: The effects of comparison publicity, performance feedback, and self-esteem. *Personality and Social Psychology Bulletin, 13*, 111–122.

Smith, S.M. (1979). Remembering in and out of context. *Journal of Experimental Psychology: Human Learning and Memory, 5*, 460–471.

Smith, S.M. (1982). Enhancement of recall using multiple environmental contexts during learning. *Memory and Cognition, 10*, 405–412.

Smith, S.M. (1984). A comparison of two techniques for reducing context-dependent forgetting. *Memory and Cognition, 12*, 477–482.

Smith, S.M., Glenberg, A., & Bjork, R.A. (1978). Environmental context and human memory. *Memory and Cognition, 6*, 342–353.

Smith, S.M., & Rothkopf, E.Z. (1987). Contextual enrichment and distribution of practice in the classroom. *Cognition and Instruction, 7*, 440–451.

Smith, W.P., Davidson, E.S., & France, A.C. (1987). Social comparison and achievement evaluation in children. In J.C. Masters & W.P. Smith (Eds.), *Social comparison, social justice, and relative deprivation* (pp. 55–81). Hillsdale, NJ: Lawrence Erlbaum Associates.

Snyder, T., & Palmer, J. (1986). *In search of the most amazing thing: Children, education, and computers*. Menlo Park, CA: Addison-Wesley.

Sorrentino, R.M., & Shepperd, B.H. (1978). Effects of affiliation-related motives on swimmers in individual versus group competition: A field experiment. *Journal of Personality and Social Psychology, 36*, 704–714.

Spear, P.S., & Armstrong, S. (1978). Effects of performance expectancies created by peer comparison as related to social reinforcement, task difficulty, and age of child. *Journal of Experimental Child Psychology, 25*, 254–266.

Spence, K.W. (1956). *Behavior theory and conditioning*. New Haven, CT: Yale University Press.

Spence, K.W., Farber, I.E., & McFann, H.H. (1956). The relation of anxiety (drive) level to performance in competitional noncompetitional paired-associates learning. *Journal of Experimental Psychology, 52*, 296–305.

Sperber, D. (1985). Anthropology and psychology: Towards an epidemiology of representations, *Man, 20*, 73–89.

Sperber, D. (1996). *La contagion des idées*. Paris: Odile Jacob.

Steiner, I.D. (1972). *Group process and productivity*. New York: AC.

Strauman, T.J. (1990). Self-guides and emotionally significant childhood memories: A study of retrieval efficiency and incidental negative emotional contents. *Journal of Personality and Social Psychology, 59*, 869–880.

Strauman, T.J. (1992). Self-guides, autobiographical memory, anxiety and dysphoria: Toward a cognitive model of vulnerability to emotional distress. *Journal of Abnormal Psychology, 101*, 87–95.

Strauman, T.J., & Higgins, E.T. (1987). Automatic activation of self-discrepancies and emotional syndromes: When cognitives structures influence affect. *Journal of Personality and Social Psychology, 53*, 1004–1014.

Strayer, D.L., & Kramer, A.F. (1990). Attentional requirements of automatic and controlled processing. *Journal of Experimental Psychology: Learning, Memory, and Cognition, 16*, 67–82.

Stroop, J.R. (1935). Studies of interference in serial verbal reactions. *Journal of Experimental Psychology, 18*, 643–662.

Suls, J.M. (1986). Notes on the occasion of social comparison theory's thirtieth birthday. *Personality and Social Psychology Bulletin, 12*, 289–296.

Suls, J.M., & Mullen, B. (1982). From the cradle to the grave: Comparison and self-evaluation across the life-span. In J.M. Suls (Ed.), *Social psychological perspectives on the self*. Hillsdale, NJ: Lawrence Erlbaum Associates.

Suls, J., & Sanders, G.S. (1982). Self-evaluation through social comparison: A developmental analysis. *Review of Personality and Social Psychology, 13*, 171–197.

Suls, J., & Wills, T.A. (1991). *Social comparison. Contemporary theory and research*. Hillsdale, NJ: Lawrence Erlbaum and Associates.

Szymanski, K., & Harkins, S.G. (1987). Social loafing and self-evaluation with a social standard. *Journal of Personality and Social Psychology, 53*, 891–897.

Tajfel, H. (1981). *Human groups and social categories*. Cambridge, UK: Cambridge University Press.

Tesser, A. (1980). Self-esteem maintenance in family dynamics. *Journal of Personality and Social Psychology, 39*, 77–91.

Tesser, A. (1986). Some effects of self-evaluation maintenance on cognition and action. In R.M. Sorrentino & E.T. Higgins (Eds.), *The handbook of motivation and cognition: Foundations of social behavior* (pp. 435–464). New York: Guilford Press.

Tesser, A. (1988). Toward a self-evaluation maintenance model of social behavior. In L. Berkowitz (Ed.), *Advances in experimental social psychology*, 21. New York: Academic Press.

Tesser, A., & Campbell, J.D. (1982). Self-evaluation maintenance and the perception of friends and strangers. *Journal of Personality, 59*, 261–279.

Tesser, A., Campbell, J., & Smith, M. (1984). Friendship choice and performance: Self-evaluation maintenance in children. *Journal of Personality and Social Psychology, 46*, 561–574.

Thomson, S.C., & Janigian, A.S. (1988). Life schemes: A framework for understanding the search of meaning. *Journal of Social and Clinical Psychology, 7*, 260–280.

Tiberghien, G. (1986). Context and cognition. *Cahiers de Psychologie Cognitive, 6*, 105–119.

Triandis, H.C. (1989). The self and social behavior in differing cultural context. *Psychological Review, 96*, 506–520.

Triplett, N. (1898). The dynamogenic factors in pacemaking and competition. *American Journal of Psychology, 9*, 507–533.

Tulving, E. (1972). Episodic and semantic memory. In E. Tulving & W. Donaldson (Eds.), *Organization of memory*. New York: Academic Press.

Tulving, E. (1983). *Elements of episodic memory*. Oxford: Oxford University Press.

Versace, R., Monteil, J.M., & Mailhot, L. (1993). Emotional states, attentional resources and cognitive activity: A preliminary study. *Perceptual and Motor Skills, 76*, 849–855.

Voss, J.F., Greene, T.R., Post, T.A., & Penner, B.C. (1983). Problem-solving skills in the social sciences. In G. Bower (Ed.), *Psychology of learning and motivation*, Vol. 17. New York: Academic Press.

Vygotsky, L.S. (1978[1930–1931]). *Mind in society: The development of higher psychological processes*. Cambridge, MA: Harvard University Press.

Waters, R.H. (1933). The specificity of knowledge of results and improvement. *Psychological Bulletin, 30*, 673–690.

Weiner, B., & Schneider, K. (1971). Drive versus cognitive theory: A reply to Boor and Harmon. *Journal of Personality and Social Psychology, 18*, 258–262.

Weinstein, R.S. (1976). Reading group membership in first grade: Teacher behaviors and pupil experience over time. *Journal of Educational Psychology, 68*, 103–116.

Weiss, R.F., & Miller, F.G. (1971). The drive theory of social facilitation. *Psychological Review, 78*, 44–57.

Weldon, E., & Gargano, G.M. (1988). Cognitive loafing: The effects of accountability and shared responsibility on cognitive effort. *Personality and Social Psychology Bulletin, 14*, 159–171.

Weldon, E., & Mustari, E.L. (1988). Felt dispensability in groups of coactors: The effects of shared responsibility and explicit anonymity on cognitive effort. *Organizational Behavior and Human Decision Processes*, *41*, 330–351.

Wheeler, L. (1991). A brief history of social comparison theory. In J. Suls & T.A. Wills (Eds.). *Social comparison. Contemporary theory and research* (pp. 3–21). Hillsdale, NJ: Lawrence Erlbaum Associates.

Wheeler, L., & Miyake, K. (1992). Social comparison in everyday life. *Journal of Personality and Social Psychology*, *62*, 760–773.

Wheeler, L., Shaver, K.G., Jones, R.A., Goethals, G.R., Cooper, J., Robinson, J.E., Gruder, C.L., & Butzine, K.K. (1969). Factors determining choice of a comparison other. *Journal of Experimental Social Psychology*, *5*, 219–232.

White, J.D., & Carlston, D.E. (1983). Consequences of schemata for attention, impression and recall in complex social interactions. *Journal of Personality and Social Psychology*, *45*, 538–549.

Wicklund, R.A., & Duval, S. (1971). Opinion change and performance facilitation as a result of objective self-awareness. *Journal of Experimental Social Psychology*, *1*, 319–342.

Wiener, N. (1948). *Cybernetics, or control and communication in the animal and the machine.* Cambridge, MA: MIT Press.

Willerman, B., Lewit, D., & Tellegen, A. (1960). Seeking and avoiding self-evaluation by working individually or in groups. In D. Wilner (Ed.), *Decisions, values, and groups, Vol. 1.* New York: Pergamon Press.

Williams, K.D., Harkins, S., & Latané, B. (1981). Identifiability as a deterrent to social loafing: Two sheering experiments. *Journal of Personality and Social Psychology*, *40*, 303–311.

Williams, K.D., & Karau, S.J. (1991). Social loafing and social compensation: The effects of expectations of co-worker performance. *Journal of Personality and Social Psychology*, *61*, 570–581.

Williams, K.D., Karau, S.J., & Bourgeois, M. (1993). Working on collective tasks: Social loafing and social compensation. In M.A. Hogg & D. Abrams (Eds.), *Group motivation: Social psychological perspectives* (pp. 130–148). New York: Harvester Wheatsheaf.

Williams, K.D., Nida, S.A, Baca, L.D., & Latané, B. (1989). Social loafing and swimming: Effects of identifiability on individual and relay performance of intercollegiate swimmers. *Basic and Applied Social Psychology*, *10*, 73–81.

Wills, T.A. (1981). Downward comparison principles in social psychology. *Psychological Bulletin*, *90*, 245–271.

Wilson, S.R., & Benner, L.A. (1971). The effects of self-esteem and situation on comparison choices during ability evaluation. *Sociometry*, *34*, 381–397.

Wine, J. (1971). Test anxiety and direction of attention. *Psychological Bulletin*, *76*, 92–104.

Wood, J.V. (1996). What is social comparison and how should we study it? *Personality and Social Psychology Bulletin*, *22*, 520–537.

Woodfield, A. (1986). *Thought and objects.* Oxford: Oxford University Press.

Zajonc, R.B. (1965). Social facilitation. *Science*, *149*, 269–274.

Zajonc, R.B. (1966). *Social psychology: An experimental approach.* Belmont, CA: Wadsworth.

Zajonc, R.B. (1980). Compresence. In P. Paulus (Ed.), *Psychology of group influence.* Hillsdale, NJ: Lawrence Erlbaum.

Zajonc, R.B. (1989). Styles of explanation in Social Psychology. *European Journal of Social Psychology*, *19*, 345–368.

Zajonc, R.B. & Sales, S.M. (1966). Social facilitation of dominant and subordinate responses. *Journal of Experimental and Social Psychology*, *2*, 160–168.

Zbrodoff, N.J., & Logan, G.D. (1986). On the autonomy of mental processes: A case study of arithmetic. *Journal of Experimental Psychology: General*, *115*, 118–130.

Author Index

Subject Index

For Product Safety Concerns and Information please contact our EU
representative GPSR@taylorandfrancis.com
Taylor & Francis Verlag GmbH, Kaufingerstraße 24, 80331 München, Germany

www.ingramcontent.com/pod-product-compliance
Lightning Source LLC
Chambersburg PA
CBHW050524270326
41926CB00015B/3051